GREEN OR GONE

David Shearman has held senior academic and medical positions at Edinburgh and Yale Universities and a Chair of Medicine in Adelaide. He is a physician, scientist, researcher and student of the environment.

Gary Sauer-Thompson is a philosopher at Flinders University who works on economic and social issues.

GREEN OR GONE

David Shearman

with
Gary Sauer-Thompson

**Wakefield
Press**

Wakefield Press
Box 2266
Kent Town
South Australia 5071

First published 1997
Reprinted 1998

Designed by Nick Stewart, design BITE, Adelaide, with David Shearman
Typeset by Clinton Ellicott, MoBros, Adelaide
Printed and bound by Hyde Park Press, Adelaide

National Library of Australia
Cataloguing-in-Publication entry

Shearman, David, 1937–   .
Green or gone.

ISBN 1 86254 426 3.

1. Environmental health. 2. Human ecology.
I. Sauer-Thompson, Gary. II. Title.

304.2

Promotion of this book has been assisted by
the South Australian Government through
Arts South Australia.

Wakefield Press thanks Wirra Wirra Vineyards

for their continued support.

contents

Preface

We fail to understand the vital importance of health to each of us. The realisation comes with disease or disability. Perhaps we have smoked and now, at the age of 40, a lung cancer is diagnosed. We can never return to the carefree days of good health. The tumour will not only shorten our life but will influence our thinking, our relationships and our values. We have slipped through a black hole into another world and can never go back. We now know what health was. But disease is conferred upon us not just by our own bad decisions, such as the decision to smoke, but by our collective community, national and worldwide decisions or by our failure to act.

This book explores the close links between health, the environment and the world's economic system. It considers human health and well-being over the next 50 years during which period we will experience worldwide changes in ecological and environmental conditions. The motivation for such a book arose out of our conviction that we – that is, you, us, and countless millions of other humans – live in ways that endanger the lives of other humans and countless species. We hold this view because when we look at ourselves and our fellow humans and at the ecosystems in which we and other species live, we come to the conclusion that by comprehensive standards we are unhealthy. Like someone with a growing cancer, we may look well to begin with, but before long, signs of impending disorder and demise declare themselves. The authors are worried that such signs are beginning to declare themselves worldwide and along with many ecological changes will adversely affect human health and that of other species.

One of us participated in a National Health and Medical Research Council of Australia (NH & MRC) working panel, whose task was to oversee the production of a report on ecological sustainability and human health. A draft report was prepared after extensive consultations.[1] NH & MRC documents are usually presented in fairly austere prose and are very much limited to what can be said with the facts at hand. But this report was different. It presented ideas that confronted many aspects of our contemporary way of life. It detailed some of the thoughts that we have explored in this book. The project was abruptly cancelled when the full committee of the NH & MRC met and reviewed the first draft of the report. Inadequate reasons were given for the cancellation. The panel accepted that some of the work was not strictly scientific and that the draft report needed significant refinement before it could have been published. But in the panel's view that was not enough to dismiss the concepts that had explicitly been sought and discussed at working party meetings. It seemed that when the going got tough, the major body charged with the responsibility of leading the nation in health matters couldn't stomach the expansive and confronting view presented to it and the implications for government, bureaucracy and our economic system.

It was this episode that prompted the authors to write a book describing the concerns we and others have about the ecological basis for our health and analysing some of the causes of the harms we observe – harms, we argue, that will need to be tackled sooner rather than later. The book is very much an amalgam of deeply held ideas of the authors, and those colleagues who have assisted them, in the areas of their day-to-day working lives. In discussing the wide dimensions of health, we range from the scientific medicine of transplantation and antibiotics, to the 'disease' of indigenous peoples – their loss of land.

The authors have backgrounds in internal medicine, biological sciences, economics and environmental issues. They have been assisted by colleagues with expertise in ecology, population biology and sociology. These different backgrounds have fostered a close scrutiny of each other's ideas and resulted in a unified appraisal of health and ecology.

As our central task, we point out the contradictions, inconsistencies and incoherencies in the ways humans inhabit the Earth and indicate how these will lead to major challenges to our health. We are not unjustly negative and whenever possible we are hopeful. We provide alternative ways of seeing the world. We envisage an existence that recognises and respects the rights of humans to have equity, but at the same time accords to other, non-human life, respect, value and recognition. Such an existence would ensure for our descendants a sustainable future, as free from suffering as possible. So it can be seen that this is not just an altruistic viewpoint, but one that is in our self-interest to pursue.

We hope this book is not read as yet another doom and gloom story. While we have reason to be pessimistic – for our failure to control liberal capitalism with its dream of a technological utopia is destroying much of nature – we have some reasons to be hopeful about our ability to live in more ecologically sustainable ways. What perplexes us and challenges our thinking is why, in the face of obvious harmful actions to nature, humans do not take a more ecologically sustainable path. Finding a useful answer to that question will occupy a modest section of this book. What is suggested is a change in individual psychology because we all need to change our conventional pursuit of self-interest. This necessary re-examining of the ethical basis of our lives involves changing our collective psychology as well. It involves questioning some of the fundamental assumptions of our cultural heritage, since the attempt to gain wealth through technological control over nature is precipitating extensive ecological damage, which will lead to devastating effects upon our health. If we are to curb industrial pollution, use natural resources more wisely and maintain and improve our health, we need to recognise the considerable structural and mental barriers in the way of change.

It is naive in the extreme to think that changing human practices, either individually or collectively, is easy or simple. Yet individuals and groups of people have changed their behaviour, sometimes in dramatic and fundamental ways – for example, in Eastern Europe when the Berlin Wall came down. That should be reason enough for us to hope that we will be able to modify our current market-oriented ways of living.

The history of the world is replete with discredited or condemned ideologies and projects. Nazism, that most horrid, vile degradation of humanity, vanished in the fires of Berlin at the end of the Second World War (though some embers have perhaps rekindled). Marx held that greed and envy are socially formed, and his ideas of an alternative to a free-market economy transformed the lives of two-thirds of the world's people; but now, in the late twentieth century, this form of socialism is largely defunct. Communist societies in Eastern Europe and the former Soviet Union collapsed because they could not give their people what they needed and because they took an authoritarian turn. The health and well-being of their peoples and environments have suffered enormously. We wonder whether a future historian will write similar histories of the collapse of consumer capitalism, which is now the dominant model for developed societies. We wonder whether liberal political democracy, now the dominant political system in the developed world, is capable of dealing with the current and future challenges described in this book.

Governments and the media see the issue in terms of economic growth versus the environment, jobs versus forests, global free market development versus environmental protection through regulation. Disagreements are especially heated over whether the market is the solution or the greatest threat to ecological health. Some believe that a civilisation built on unlimited economic growth through treating nature as tap and drain can keep on with 'business as usual' forever. The depletion of the ozone layer, which shields our planet from solar radiation and global warming, is evidence of the implausibility of that assumption.

We have some utopian visions. We wonder what a world

would be like if most people had an effective understanding of how to live in an ecologically sustainable way. What would the political system be like? What would everyday life be like, and how would human relationships be altered? We suggest that it would be a different world from the one in which we now live; it would be one based on a dissatisfaction with the basic structure and direction of our present society. The welfare of the human and non-human inhabitants would be the central concern. We predict and hope that concentration on ecological living will be the dominant method of change in the coming years. We recognise that unless we look after each other we are not likely to look after much else. We have faith in the ability of humans to alter the course of their lives, as individuals, as communities, as institutions and as political systems. We see people designing strategies for the future that will lead to better health for all humans and better health for ecosystems.

In developing the themes of this book, we emphasise that we are not naive in the sense that we hark back to some golden age, a Green utopia where we live in harmony with our surroundings. It is doubtful if humans did so comprehensively in the past, and it is unlikely that we could do so in the future. Our reproductive potential has been amplified so that we affect other organisms and environments in ways that no other species can. But the consciousness that accompanied our evolution also means we have the capacity to reflect upon and change our behaviour.

But having utopian visions and designing strategies will not be enough. As Allen Wheelis has said in *How People Change*, you don't become a thief by taking an item from a store once in a while, you become a thief by working at it, by stealing and stealing again until one morning you wake up and you are a thief.[2] We will not become ecologically responsible humans by joining an environmental group (though this is an important start) or recycling occassionally or by saving part of a rainforest, while we merrily go about exporting our top-soil and clearing vast tracts of native vegetation or giving our young people little hope of gainful employment. We will only become ecological humans, that is ecologically virtuous or caring humans, by changing our identities and character through explicit, repeated actions that are consistent with ways of living that are sensitive to our own needs and those of non-human life.

The two authors and their colleagues who have assisted are Australian. This could be seen as a disadvantage when the book is intended to have a global perspective, but we believe it is actually a strength. There is a view that if Australia – spacious, wealthy, democratic and successfully multicultural – cannot solve its own environmental and health problems and show leadership to the world, then what hope is there for those countries facing poverty, racial dissent or dictatorship? The reader will find, therefore, that many of the examples about health and ecology are Australian ones. The intent of this book is to encapsulate all relevant issues in a straightforward way for the general reader, the relevant issues being not only individual and population health, and global warming, but our psychology, value systems, philosophies, economics, and politics. All these are linked in the creation of our health problems and in the ways that we must solve them.

Recently, after prolonged drought, huge fires burned across the tropical rainforests of Indonesia. Much of South-East Asia was enveloped in acrid smoke. It is thought that global warming had accentuated the El Niño ocean current which causes drought. The fires started with logging and burning of residues, and clearing of land for plantations. Government and industry displayed apathy, ignorance and greed. Fires continued to be lit while firefighters from other countries struggled to extinguish them. A prominent tycoon and logger blamed the indigenous peoples and said that the environmentalists who complained were communists! The fires were a global catastrophy for greenhouse gases poured into the atmosphere and land and biodiversity was lost. Your grandchildren will suffer the health consequences. This book explains why.

1. 'On Which All Life Depends: Principles for an Ecologically
   Sustainable Basis for Health', NH & MRC, December 1994.

2. Allen Wheelis, *How People Change* (Harper & Row, New York, 1975).

CHAPTER 1

WHAT IS HEALTH?

Adelaide, where the authors live, is a small city in the state of South Australia. It is a city of one million inhabitants in a state of one-and-a-half million. Beyond this picturesque city are urban market gardens and beyond them vast agricultural and pastoral lands with grain a major crop. Aboriginal people inhabited South Australia for thousands of years and modified the original vegetation by using fire to flush out animals when they hunted and to encourage regeneration of plant foods. European settlement commenced in 1836 and the land was cleared aggressively. The capacity of the land to support the native animals and vegetation was greatly reduced. By the 1990s less than five per cent of the landscape could be described as 'bush' or largely unchanged native vegetation. Exotic plants like broom and blackberry and animals like the domestic cat, the domestic mouse, the fox and the rabbit have spread throughout the land.

In 1993 a plague of mice occurred in an agricultural region about 200 kilometres from Adelaide. By all accounts it was a severe plague. Mice in their millions were practically everywhere – in kitchens and beds, in granaries and fields. No pied piper could control this massive plague. Its origin lay in the imbalance of species. The normal interaction between many different species was changed to such a degree that the system as a whole destabilised, with drastic consequences: the conditions and food required for a huge expansion of the mouse population occurred. A superabundance of grain, adequate burrowing sites in ploughed land, mild weather and absence of competitors and predators caused the population explosion. The plague might have been described simply as an inconvenience, were it not for the destruction of grain stores and of electrical wiring by gnawing mice, which caused fires.

The population expanded and moved into the outskirts of urban Adelaide where protection for the mice was greater and food abundant. Large-scale eradication measures, such as strychnine, which had been used in the rural areas, were no longer feasible – the danger to humans and to domestic and other animals was too great to use it. The usual mouse poisons such as warfarin were too expensive to use. And so these small creatures defecated and urinated everywhere. No cup or plate was secure. Exposure of humans to the excrement of mice on a large scale was inevitable under these circumstances.

Fortunately no diseases were transmitted to humans but such plagues of rodents are a significant threat to human health. Some rodents carry viruses called hantaviruses which are passed in their urine, faeces and saliva. During a plague of mice in south-western USA in 1993, 124 young people developed infection of the lungs which suffocated and killed half of them. A new strain of the hantavirus carried by the mice was responsible. In both South Australia and south-western USA, disturbance of the environment by humans was responsible for the mouse plague.

These events have similarities to the disease the Plague or the Black Death, which wreaked havoc in Europe in the Middle Ages. The word 'black' comes from early descriptions of the disease which described swellings in the groin and in the armpits which rapidly increased in size to that of an egg or an apple. Black blotches then developed over the body and these rapidly increased in size. The tongue and throat became swollen with blood and turned black; in some cases small black blisters spread widely over the body. Death was horrible. The causative germ was the bacillus *Yersinia pestis*, named after the word 'pestilence' which meant a plague or other deadly epidemic. The Black Death existed in epidemics for many centuries killing millions of people as it swept across Asia and

Europe on several occasions. It is thought that the nursery rhyme 'Ring-a-ring o' roses' had its origin in the Black Death.

Ring-a-ring o' roses,
A pocket full of posies,
A-tishoo! A-tishoo!
We all fall down

The ring of lesions on the skin was an early sign of the Plague. Posies were carried as a protection. Sneezing was the final fatal symptom and the infected person then 'fell down' dead.

In this book we will describe how a disease affects the average person – a 'case report' as it might be described by a medical doctor. Some of these 'case reports' are based on real case studies but the names and places are changed. We commence with an historical study.

Bessy Bell and Mary Gray were two young girls living in Perth, Scotland in the seventeenth century. The Black Death swept into Perth from a trading ship which had anchored in the Firth of Forth. Death was quick but terrible and the carts collected hundreds of bodies from streets and homes every day. Bessy and Mary made a pact to run away into the country to escape the plague. Frequently the population of cities scattered in this way in the hope of avoiding death. The friends ran to a place called Burn-braes outside Perth where they constructed a small bower from branches, wood and leaves, to give themselves shelter. Two days later they became breathless and died in each other's arms from suffocation. A memorial marks the spot. Bessy and Mary had inhaled the germs of Black Death before leaving Perth. These had been transmitted to them in droplets from the coughing of other infected persons. They had died from pneumonic plague.

In Chapter 2 we explain ecology and ecological diseases, but as a simple introduction, an ecological disease is one caused by a disturbance or imbalance of species, be they plants, animals, bacteria or viruses. The Black Death has many of the hallmarks of an ecological disease. The germ responsible, the bacterium *Yersinia pestis*, occurs in many rodents throughout the world. It infects these rodents without them dying and is transmitted between them by fleas and by direct contact. These rats develop immunity to the plague simply because they live with it. Pockets of infected rats exist to this day in many parts of the world in a stable relationship with their infected fleas.

What changed this stable situation into the epidemics of plague? One theory is that rapid expansion of the human population in parts of Asia and Europe led to the clearing of land for farming and the crowding of people into already over-populated towns and cities. People lived in unhygienic conditions and in very close contact. In the fourteenth century it was thought that a period of unusually cold weather together with crop failures led to rats forsaking the granaries and fields, to invade the towns and human dwellings. These rats were the species *Rattus rattus*, the common black or brown house rat, which had not been exposed much to *Yersinia pestis* and did not have immunity to it. When infected, this rat, like the human, died rapidly of overwhelming infection. The fleas jumped onto humans before or after the rat's death and so transferred the infection. Moreover, in the fourteenth century there was great activity in the trade routes between Asia and Europe both by land and by sea across the Black Sea, the Mediterranean and the North Sea. Often the disease appeared first in ports and then spread inland. The ships carried colonies of diseased rats and it is possible that sick rats or fleas travelled in the goods that were being transported across land. The Black Death was an ecological disease. Its spread depended upon changes in the relationship between rat, flea and man brought about by land clearance and the crowding of humans into urban areas.

Our revulsion of the rat is a reflection on its habits and its carriage of many diseases. Man and rat are by far the two most successful warm-blooded animals on Earth. In recent history both have undergone a population explosion. Both are intelligent and adaptable to all climates, and are environmentally destructive. Rats can search out any food and they will gnaw through anything to get it – even concrete! We have no effective measures of eradication. Rats know if food is poisoned; some die from the poison but the knowledge of a laced food is transmitted to others. Millions of rats inhabit the cities of the world and they will be a major health hazard in the burgeoning cities of developing countries with their shanties and squalor, for rats will eat anything, including faeces.

Plague transmitted by rats exists in the world today. About 3000 human infections per year are recorded by the World Health Organisation, mainly in the developing countries but a few have occurred in the United States. Rats also transmit leptospirosis to humans. This bacterial infection is excreted in rat urine then penetrates the skin of humans who walk in infected water or mud and splash themselves. But it is perhaps more worrying that rats and mice carry a large number of viruses such as the hantaviruses which infect humans. The mouse is also an important link in Lyme disease (described in Chapter 2).

There are some similarities between the Black Death and the present-day spread of the HIV virus which causes AIDS. This disease may have ecological causes. The HIV virus undoubtedly existed in groups of apes and monkeys in Africa for perhaps prolonged periods of time before it spread to humans in those areas. The disease may not have spread further but for the rapid displacements of populations in Africa which was caused by the replacement of tribal structures with Western rule and economics, and but for our modern ability to travel vast distances by air routes and infect members of other populations by sexual contact. HIV infection is discussed in more detail in Chapter 3.

We will explain how our destruction and mismanagement of the Earth's environment may affect our future health and – if we do not heed the warnings – the future of civilisation as we know it today. We have set the scene by explaining how complex relationships between humans, animals and infections in the environment can lead to catastrophic epidemics of disease. Before we go further we must discuss the meaning of health so that we understand all the factors that will allow us and future humans to live a healthy life.

People in Western society think of health as a long and fulfilling life without disease. Our fears are of cancer, heart disease, stroke and disability. When disease occurs scientific medicine restores health. By contrast, health to the squatter in Rwanda means cessation of war, for war brings not only violent death and injury, but also famine and disease. To the millions of poor in Bangladesh, health means clean water and adequate nutrition. Health to the Aborigine in Australia is tribal living on their inherited land. So a definition of health needs to take into account the values of different societies.

Health in our Western society is dominated by the cure of disease by medical science. Indeed a 'health service' is mainly a 'disease service' which is very expensive. The role of the doctor is to fight a war against disease, by which we mean any condition that invades the body or stops it working properly. The main task of the doctor is to diagnose the disease by interpreting the patient's complaints and symptoms and by looking for abnormalities. Blood tests, X-rays and scans are used to confirm the diagnosis. The doctor aims to restore the body to normal biological functioning. There is no doubt that medical science has achieved miracles resulting in longer and healthier lives, especially for those in Western society. Medicine has been successful beyond all expectations. The dreaded leukaemias can be treated and in most instances cured, and gene therapies which insert genetic material into cells promise to eradicate some of our most feared genetic diseases such as cystic fibrosis. Whilst the chronic diseases and disabilities have been more difficult to help, joint replacements, particularly of the hip, have alleviated pain and suffering and allowed thousands to walk again; peptic ulcer can now be cured in most cases, for it has been found to be due to an infection of the stomach by an organism called *Helicobacter pylori*. Twenty years ago, thousands of patients were still undergoing surgery to the stomach to ease their pain and suffering. Huge advances have been made in what is called supportive therapy – resuscitation and keeping the accident victim alive until surgery can be undertaken. Many lives can be saved by the transplantation of kidneys, hearts, lungs, livers and bone marrow.

In 1962, in the Royal Infirmary of Edinburgh, Scotland a young intern, qualified only a few months was awake at 3.00 am with panic. At 6.00 am a surgical team was assembled in the showers adjacent to the third floor surgical theatre. The ablutions were long and thorough as every possible germ was scrubbed from the surface of their bodies. For the first time a kidney transplant between relatives was about to begin. Although the press hovered, secrecy was maintained and at 6.30 am the team, pristine-clean, shuffled into the 'space capsule' of the operating theatre for an historic 12-hour journey. Tension crackled in the air. The leader of the team, the pioneer, an eminent surgeon born in London, England, fussed and complained about the instruments, the swab count, the blood vessels from the donor kidney (removed by different surgeons in an adjoining theatre) which had been cut too short! The young intern had to hold onto a deaver (a bent piece of stainless steel used to keep open the wound) for hours without moving. Each hour a nurse with a drink of orange juice would carefully ply a straw around the face mask of each member of the team. The young intern, bent on personal survival in a competitive profession, realised that he must drink enough to prevent dehydration, but not too much in case he had to pee! History was made: the patient and the donor survived as did thousands of others to follow. The young intern is one of the authors of this book and the experience brings home the vast technological and biological advances made in one professional lifetime. Transplantation is now a routine procedure with universally excellent results. The future could not have been imagined in the panic of that day in 1962. Then, the sceptics thought of the operation as heroic but bound to fail.

There promises to be no end to the biochemical and technological advances in scientific medicine, but scientific medicine has limits. It often fails to deal with what we may call quality of life, for example the difficulties after an accident that causes the loss of a leg or paralysis. While the loss of movement may be compensated for by a supportive family, there is little to be done about the sadness and ill-health that is constituted by the loss. Indeed our Western societies measure the economic value of this loss. Financial compensation is decided by the social and economic values of our society. A promising young male film star, severely injured in a car crash, will be awarded millions of dollars in our courts. A mother with three young children to nurture will be lucky to get more than the tiniest fraction of such an award. That the film star's illness is seen to be more important than the mother's has to do with our values, economics and politics, not medicine.

Ian Kennedy in his book *The Unmasking of Medicine* argues that the doctor is seen as a mechanic, the patient as a machine or a car, and illness as a mechanical failure.[1] Just as a car is taken to specialists in transmission, body work, gear boxes or exhaust systems, medical science concentrates on the diseased or malfunctioning part of a human machine. Scientific medicine does not see the sick person as a whole, in the physical, social and ecological environment of which he or she is a part. This situation is well recognised in the gossip of the young hospital doctors: 'Old so and so (the surgeon) was delighted to hear that the eye surgery was so successful – pity that the patient died.'

Scientific medicine concentrates on 'real' diseases, those that can be diagnosed and cured by science. But it avoids the psychosocial issues which are not thought of as 'real diseases' as they cannot be explained scientifically. It does not recognise that disease can be a disease of industrial or modern civilisation. Social and cultural pressures cause to varying degrees obesity, anorexia nervosa, alcoholism, depression, drug dependence and excessive gambling and indeed smoking. These are diseases for they impair health, happiness and well-being. Scientific medicine certainly recognises the results of excessive alcohol and of smoking. Excessive alcohol causes cirrhosis of the liver, torrential bleeding from the stomach so that blood is vomited, and damage to the heart and brain. But should the excessive alcohol consumption that leads to these disasters be called a disease? We think so. Scientific medicine does not.

Scientific medicine often acts as though symptoms are only meaningful if their investigation leads to a clear diagnosis. But many disorders show no abnormalities when investigated. A patient with constant, debilitating, abdominal pain may undergo every investigation known to medical science without any abnormality being demonstrated. Such conditions are labelled 'functional' because the physician reasons that there must be a 'functional' disorder of nerves or bowel muscles which is causing the pain but which we do not yet detect or understand. Scientific medicine responds to this challenge with research programmes on the movement of the intestines (bowels) and its control by the nervous system. Often, however, careful exploration of all the patient's circumstances may reveal an overwhelming personal problem. The patient may have a partner who is addicted to gambling, so placing the family in constant debt. Another may be bullied or sexually harassed at work or still be grieving over the death of a child many years previously. Indeed, many patients with pain do not have any recognisable disease. They may have headache, abdominal pain or chest pain arising from stress, social malaise, alienation or depression. Western medicine does not *cure* these patients, though some doctors have the skills, insight and sympathy to help them.

Yet scientific medicine with its view of health as the absence of disease is dominant in our culture. Its prestige, power and access to funding have resulted in a public fascination with disease, deformity and injury and the belief in 'cure' by surgery or medicines. The public demands and finances these 'cures'. The pharmaceutical industry will spend millions of dollars developing a single drug, knowing that if it works there will be huge profits. The wealthy Western public donates fortunes to research into heart disease and cancer – to some extent diseases of Western lifestyle – but scarcely a cent to investigate malaria or enteric infections, which kill millions of children in poor countries. A *cure* for HIV infection, the cause of AIDS, is a public commitment of governments and presidents, yet prevention remains controversial and its funding is poor or non-existent in the countries where the greatest proportion of people are infected.

We need to question scientific medicine's view of health. Action and money are for individual 'health care' (i.e. cure of a disease in the individual) rather than for the prevention of disease by public health measures. In Australia in 1993, the National Health and Medical Research Council spent ten times as much money on scientific research related to disease than on public health research.

It also has to be recognised that the health of most of the world's population in terms of well-being, life-span, infant and maternal mortality and the conquest of common infectious diseases, has not been improved significantly by cures offered by highly technical Western medicine and surgery. Rather, it has been improved by public health – preventative measures which provide clean water and sanitation, adequate nutrition, access to basic health care and family planning to millions of people. Furthermore, Western scientific medicine has become very expensive. In many countries 'economic rationalism' has categorised expenditure on health as non-productive; budgets have been cut and the standards of care in public hospitals have fallen. The trend is that Western scientific medicine is for those who can afford it; the poor are being increasingly deprived of good basic health care. We conclude that scientific medicine, despite its many achievements, does not give us a healthy life. A healthy life depends upon a wide spectrum of factors which we discuss below.

Western definitions of health

The answer to what it means to live a healthy life may seem obvious – freedom from disease, and long life and happiness. In other ways the answer is more complicated: we can never avoid disease whether infectious or other, and happiness is elusive – it has changed in its meaning from living a good human life, to individual pleasure or contentment. Attempts to define health have never been truly successful. Indeed, we should not expect a tight definition of health for health means different things to different individuals, communities and cultures. However, when we consider populations, we do have methods to define health. The health of a population can be measured by how long people live on average and by how many people are ill or are dying. Such measurements tell us that people in Western countries are on average healthier than those in developing countries. They also tell us that world-wide there have been definite improvements in life-span.

We live in a world dominated by Western scientific medicine and continue to think of health as an absence of disease but leaders in the health professions have moved to a more social meaning for health. The World Health Organisation (WHO) brings together health experts from many countries to prepare policies for worldwide health, to set standards and – if possible – to act when health is endangered. When it was founded in 1948 the WHO provided us with a more social definition of health: 'a state of complete physical, mental and social well-being, not merely the absence of disease or infirmity'.

The introduction of the words 'physical, mental and social

well-being' was revolutionary. It recognised that health is the product of political, social and economic conditions. Illness could be genetic, or due to infection – but it could also be the product of today's culture. The judgement that someone is ill is therefore a value judgement and not simply a statement about the symptoms of the malfunctioning body. The values that doctors refer to in making their judgements about what is illness and what is not, are those of their society and those accepted by society as a whole. If the state of normal functioning – that is, health – is living a human life well in a particular human community, then well-being needs to be linked to economic development.

The problem with the 1948 WHO definition of health is that 'well-being' can be interpreted as the desires of each individual. This is especially so in affluent countries such as Australia and the USA. For most people in these countries, well-being is experienced sitting in front of the television eating a good dinner, after driving home in a new car from a hard day of money-making.

But in 1948 in a world shattered by war, 'well-being' probably meant food, shelter, employment, peace and freedom. Unfortunately this was not spelled out at the time, but since then the experts in WHO and elsewhere worldwide have used the words, food, shelter and so forth to describe 'well-being' and we now accept that they are necessary for health. Furthermore, for most of us well-being would mean having a fulfilling life, being educated within our capabilities, having a job, achieving our goals to our satisfaction and being part of society with friends, family and community. All these issues are linked to health.

Consider unemployment for example. To sit on the sidelines watching others work leads to loss of esteem and a feeling of worthlessness. The unemployed become depressed, make bad decisions, suffer poor nutrition, lose friends, and are tempted into petty crime – perhaps only the shoplifting of a chocolate bar thrust at them daily by the television advertisements of our consumer society. Is unemployment a disease? You may say no, because it is not included in the medical textbooks but it *is* a health hazard – the unemployed become much more likely to have both mental and physical illness. Therefore, to achieve the food, shelter, employment and participation in society that all constitute 'well-being', we need resources.

The 1948 WHO idea of health believed that economic growth would provide those resources we needed to be healthy in the broadest sense. Economic growth would bring jobs for most, money, and access to a public health system. Well-being would follow. But will economic growth continue to result in well-being forever? Some of us already know it will not. There is a growing realisation of the conflict between growth and the environment and that our well-being is under threat from ecological disturbances. This recognition of environmental issues was reflected in the wider official understanding of health voiced by H. Mahler, a former director-general of WHO:

> *We are now witnessing that the term physical well-being means much more than the biology of the human body: it includes a safe environment and the responsibility for our physical surroundings on the planet as a whole.*[2]

A responsibility for our physical surroundings implies a recognition of economic growth as a cause of environmental damage and illness. It became necessary to take responsibility for our actions towards nature, by preventing its destruction and by ensuring that the damage to our environment does not destroy human well-being. It implies that living purely for economic growth and wealth creation for its own sake would be living badly. Today we put environment second to growth, whereas for the future health of the human race we have to learn to place the environment before economic growth. The present degradation of the environment suggests that to achieve well-being within a capitalist market economy, it is necessary to take political action to regulate environmental degradation. As we shall see in later chapters, we are caught in a quandary because slower economic growth is politically unacceptable because of the perceived need to be internationally competitive.

Some of us now recognise that the concept of health needs to be broadened to include ecologically sustainable development and a consideration of all other life worldwide. Sustainability requires managing resources so that they remain available and unchanged for future generations. We are coming to realise that economic growth must be subordinated to the preservation of and the health of the environment; a free-for-all market economy is becoming increasingly incompatible with sustainability.

In the 1990s at the Rio Earth Summit it was at last recognised that human health had to be seen in the context of the total human environment, including the cultural and economic environment, and the health of the physical and living world itself (the biosphere). It was seen that there was a deep conflict between development, human well-being and the biosphere, evidenced by acid rain, global warming and ozone depletion, caused by an economy that is to grow apparently forever. It was recognised that health could not be defined narrowly as an absence of disease, for it involves our relationship with the land and the way we dwell on the Earth. Our present way of life is destructive to the land, waterways, soils and biodiversity.

Health and the land

So far, we have explained that health is not just the absence of disease. Health depends on adequate food, shelter, employment, friends and society. And there is an even wider dimension to health. The health of each of us depends upon our relationship to our environment and even more importantly our ecology – our relation-ships with the non-human world. Yet we are disturbing and destroying these other living things. Poor health is often a result of our present way of life which uses all nature as a resource for economic growth and disturbs the ecological system, thereby causing ecological diseases.

In marked contrast to the Western attitude, indigenous peoples in Australia and New Zealand, the Aborigines and Maoris, and indigenous peoples elsewhere do understand health in terms of the relationship of their society to the environment. Their concept of health and well-being was based on their care for the land, because it was part of their identity as a people. They developed what would now be called a 'land ethic' in which they recognised that the natural environment had value in itself. Their community included soils, waters, plants and animals, or collectively the land. It was part of them. The land was an object of moral concern and they therefore differed from the European invaders who saw the land as empty and unproductive until it could be turned into farms and cities.

We regard land as property, we conquer it and shape it to our will. Today in Australia our concerns extend only as far as the Land Care organisation, to conserve *resources*. There is little in the way of a land ethic based upon the viewpoint that we are dependent on all other forms of life. The land ethic of indigenous peoples integrates environmental and economic concerns. It preserves the human species but not at the expense of the other life and the land. It is important for us to understand this indigenous viewpoint for it can help us develop a different way of living on the land instead of treating it as a property and resource to exploit. Its moral perspective can help us to understand health and well-being.

Maori concepts of health

Like many other indigenous people, the Maoris were dispossessed of their land by colonisation. This disrupted their culture and lifestyle, with profound effects on their health and well-being. But they retained their traditional stories, which contain their views on the environment. To the Maoris, a healthy and well adapted person is one who makes use of the resources of the land but also respects the environment in which he or she lives. They believe that all things have a life, character or place of their own, which must be respected. In other words, they respect the environment.

Maori people believe in four cornerstones of their health. Just as each of the four sides of a house is necessary for strength and symmetry, the spiritual, mental/emotional, family/community, and physical aspects are all needed for health. As explained by Mason Durie in the book *Whaiora, Maori Health and Development*, the Maoris understood the links between humans and the environment.[3] Their faith was both a belief in God and a relationship with the environment so that parts of the landscape have spiritual significance quite apart from their economic or agricultural value. Indeed tribal elders regarded a lack of access to tribal lands as a sure sign of poor health because the natural environment was part of their identity and well-being.

Traditional Maori society had a holistic view of health which Western societies are starting to understand for themselves today. By 'holistic' we mean that they understood instinctively all the factors that we discuss in this chapter. Their holistic view of health also included their extended family, inherited strengths, and their relationship to ancestors. In many ways the Maoris had a complex and supportive culture which maintained health through ritual, laws and diet. The significance of a clean environment for good health was recognised. The risk of the transmission of infection was reduced by widely separating eating places from places of defecation. The disposal of human waste by the colonisers into areas where food was gathered, such as streams and lakes, greatly offended them. One is left with a sense of shame at the impact of the British riffraff who, used to drinking faecally contaminated water from the River Thames, then subjugated a society that in many ways was more socially and culturally advanced. The settlers had an economic system that relied on greed and a false sense of cultural superiority. Dr Isaac Featherstone, a surgeon, member of the House of Representatives and superintendent of Wellington Province in the early days of the colony, said it all:

> *A barbarous and coloured race must inevitably die out by mere contact with the civilised white; our business therefore and all we can do is to smooth the pillow of the dying Maori race.*[4]

In contrast to these colonial attitudes the Maoris had a perspective on the environment, health and well-being that we can use today. As with many indigenous people, the health of the individual and the group rests not only on relationships among themselves but on their relationship to the land, which is part of their spirituality and their concept of a good and valuable life. The theft of land from indigenous peoples not only makes them poor and more susceptible to diseases that flourish under the conditions of poverty, overcrowding and malnutrition; their society and culture are so severely harmed that they often lack the will to assist in the retrieval of their everyday health needs.

Aboriginal concepts of health

In the case of the Australian Aboriginal people, their close relationship to their land underpins their existence and indeed it is increasingly, though belatedly, recognised by Western cultures that it is necessary for their survival and identity. The loss of land to a tribal Aborigine would be as devastating as paraplegia after an accident to a person of Western culture. Neither would feel themselves to be a complete person again. That Aboriginal culture requires ownership and management of tribal land was the basis for the far-sighted Mabo decision on Aboriginal land rights. Tribal land was necessary for hunter-gathering which provided a healthy diet to which the Aboriginal peoples were adapted. On a Western diet they suffer obesity, diabetes and many other medical conditions that shorten their life-span. The land was also important spiritually as part of the 'dreaming' that gave them a sense of oneness with ancestors and the environment. Health was damaged if this way of life was disrupted. Health also depended upon social rules, relationships and rituals. Disease was considered in the same light as 'troubles' such as family fights which prevented 'good health'. When ill health occurred, the approach to a cure recognised personal and spiritual needs as well as physical ones and the return to good health depended on an ancestral relationship to the land.

Relevance of indigenous concepts of health

Today we see and hear much of indigenous peoples in documentaries, articles and books. This is because economic greed continues to drive them off their land in Malaysia to build dams, in Brazil to clear land for mining and logging and in Papua New Guinea where rivers have been poisoned by tailings from mines. The responses and statements of the indigenous peoples to these catastrophes are remarkably similar. Their life with the land, the forest, and the river is happy. They cannot leave their ancestors. Indigenous peoples are adapted to and are part of their environment; they have achieved the sustainability which Western culture talks about but does not act upon.

We are not idealising the health of indigenous peoples, nor do we wish to deny science in favour of religion. Rather, we are advocating learning from these experiences. Populations in the past, as today, varied enormously in their physical healthiness – humans can compensate for hostile environments but only so far. In general, the physical health of hunter-gatherer populations was likely to have been inferior to the health of Western peoples (with some exceptions). However, there were possibly hundreds of thousands of years during which many of these communities lived in harmony with their environment.

The Maori and Aboriginal ideas of health had a more holistic perspective than that of scientific medical and public (population) health. In particular they had an ecological idea of health from which we can learn. The individual in Western society has conflict between the pursuit of self-interest and the needs of others and the environment. The Maoris, however, had a way of living which emphasised impersonal duty, contracts, rights, and the calculation of costs and benefits. By contrast the imperfect economic and political aspects of our society place economic growth on a collision course with the environment. We can learn from indigenous peoples that well-being is tied to a relationship with nature and that we need to care for the ecosystem in which we live rather than treat it as a resource for profit.

Public health (also called population or environmental health)

So far, we have seen that health is much more than the absence of disease. The World Health Organisation definition of health as 'a state of complete physical, mental and social well-being, not merely the absence of disease or infirmity' is inadequate to explain what it means to lead a healthy life. It gives us a concept of health in which adequate food and shelter are fundamental constituents, together with certain minimum conditions and resources which enable us to survive, reproduce and function as individuals and as populations. It does not give us an ecological concept of health. This new horizon for health is wider than the present concept of public or population health. We will now describe public (population) health, and ecological health.

As humans evolved over thousands of years, they initially suffered infections and nutritional deficiencies. More recently environmental pollutions and Western lifestyles causing, for example over-nutrition, have caused disease. All these acted *locally* in regions or countries and have a simple cause-and-effect sequence. The following is an example of an environmental (public) health problem. We can find the cause which leads directly to the disease.

John Ivanovic brought his family to Australia from Yugoslavia in a search for a better life. He had skills in engineering and accepted work at a mine in Western Australia. Although isolated, his family settled well in the small mining town, which offered them housing, a good wage and a swimming pool. John worked in or near the mine for only five years. His wife wanted more contact with immigrant Yugoslavs elsewhere in Australia and they were worried about schooling for the children. They moved to Melbourne where John worked as a water engineer. After 15 years in Melbourne he became quite breathless over two days. There was a dull pain over his ribs and the local doctor did an X-ray of his chest. 'There is a lot of fluid on your right lung and this is why you are breathless; I want you to see a specialist' was the verdict from the doctor. A week later he was admitted to a large public hospital in Melbourne, a tube was pushed into the right side of his chest and the fluid was drained. It was stained with blood and contained cells that looked malignant. Now that the fluid was drained the X-rays showed that the lung was compressed by a large tumour. It was a mesothelioma, a tumour of the lining of the lung. John had worked at an asbestos mine and his tumour was caused by the inhalation of asbestos fibres. John's pain and breathlessness became intolerable and he lived for only a further six months. There is no treatment for his condition.

John's wife and three daughters had all been exposed to asbestos as well. In Western Australia the tailings from the mine had been used to pave the driveway and the road outside their house. On the frequent dry, windy days, they breathed asbestos dust. They will carry the risk of developing mesothelioma for the rest of their lives. Lung disease caused by asbestos is a severe and terrible disease. Asbestos exposure has occurred in mining and in the building trade, where asbestos was used for insulation. Public health identifies such health risks and aims to eradicate them. We can show that asbestos causes mesothelioma and we can eradicate the risk by abandoning the mining and use of asbestos.

So population or environmental health employs prevention at a community level. It includes measures to provide clean water and prevent cholera and other infections of the bowel. Sanitation prevents these infections, which are caused by human excreta entering the water supply. Clean-air regulations prevent pollution which causes severe respiratory diseases. Childhood vaccination prevents death and disease from many common infections such as measles, whooping cough and poliomyelitis. In recent times industrial pollution has been recognised and prevented if it suits governments to do so economically. The management of passive smoking (causing lung cancer) and environmental contamination with lead (causing mental deficits in the young) all fall within the population health arena.

In the sphere of nutrition, population health measures fortify our food and water with minerals to prevent diseases, for example fluoride to prevent caries in Western communities and iodised salt to prevent hypothyroidism and goitre (swelling of the thyroid gland at the front of the neck) in the iodine deficient regions of the world. This also reduces mental illness and intellectual impairment in these populations. Indeed, iodine deficiency is the most common preventable cause of brain damage in the world today. Over 20 million people are intellectually disabled because their mothers suffered from iodine deficiency during pregnancy. Iodination of salt eliminates the problem. It is a simple solution but difficult to deliver to poor and remote communities. The International Council for the Control of Iodine Deficiency Disorders, an organisation based in Adelaide, Australia, provides a structure for liaison with industry and government and for education and delivery of programs.[5]

While population health has had significant impacts worldwide there are nevertheless great differences in basic health conditions between the developing and the developed world. These differences are due mainly to the inequitable application of population health measures. Thus, serious childhood infections are eradicated by vaccination in the developed world, but poverty stops the purchase of vaccines by many developing countries. Half of the people born in the developing world today will be dead by the age of 40 years. In Australia, half will be dead by the age of 74 years. In the developing world almost 50 per cent of deaths are due to infectious diseases and events in childbirth or early life (compared with less than five per cent in a country such as Australia). The infectious diseases responsible include the enteric infections (gastro-enteritis), caused by poor sanitation which leads to contamination of water with faeces. They account for millions of deaths that are preventable by simple population health measures.

Paradoxically, in some parts of the developing world there are states or provinces where, unlike surrounding areas, the gains in health compare favourably with the Western world. Examples are Kerala in India, and Sri Lanka. The key factor seems to be that in these regions there is a more equitable distribution of basic resources within the population. This distribution is ensured by political systems in which the power of women at the household level is considerable. Even within the developed world there are great differences in life-span and illness. In some situations pockets of the Third World exist; the health of some Aboriginal Australians is a good example. In Australia the life expectancy of Aboriginals born in 1984–1989 is about 59 years compared to 76 years for white Australians. For Afro-Americans living in Harlem, adult male survival is less than for men in Bangladesh.[6] In these populations there are extremes of poverty and violence. In some situations disease patterns resemble those of the Third World with a high infant mortality, for example, among Australian Aborigines the rate is 30 deaths per 1000 births compared to 1 per 1000 for white Australians. Even so, the rate of 30 is much less than the low income countries of the developing world with 71 deaths per 1000 births. On the other hand the causes of premature death in countries such as Australia are diseases of lifestyle including heart disease and stroke, cancers, accidents, injuries (including suicides), and as a result of the use of tobacco, alcohol and illicit drugs. There are higher rates of mortality and illness among the poor and the less well educated. Relative poverty is associated with poor health from many diseases in both rich and poor countries.

In the developed world the challenges for population health are to reduce relative poverty while attending to lifestyle diseases caused by social circumstances. On the other hand, in the developing world the abolition of absolute and relative poverty required for health is most directly linked to the environment, with ecological and human impoverishment sitting side by side. By contrast, in the developed world, we feel ourselves to be divorced from the environment and this creates an illusion of independence from environmental issues. This illusion enables us to lead fairly extravagant lives as individuals and populations.

Ecological health

Population health is about local cause-and-effect problems. The drinking water in a certain area is contaminated with cholera infection and this will be corrected by establishing sanitation in the relevant villages. The mining of asbestos is so dangerous to human health that it is stopped and substitutes for asbestos are found. By contrast, ecological health has an extra dimension, usually a global one – it is concerned with the health consequences of large-scale disruptions of nature due to population growth and our consumption of energy and materials. The disruption of complex natural systems leads to a variety of scenarios which are themselves complex and which may not occur for decades. Most importantly, the end results cannot be predicted with certainty in the same way that we can predict that poor sanitation will contaminate water and lead to enteric infections including cholera, or that asbestos will damage our lungs resulting in asbestosis or mesothelioma, which is a tumour of the lining of the lung. It is easy to see that our ability to deliver individual health care is dependent upon the basics of environmental health – clean water and air, adequate nutrition, and so forth. But it is more difficult to see that all our advances and achievements to date are also dependent upon our recognition and implementation of ecological health. Nature ensures that we have a stable climate, shields us from ultraviolet radiation, and provides us with food and water and the biodiversity to ensure that food supplies are sustained. When humans disrupt these systems, the health of the world's population is endangered. Our ecological health is at risk and ecological diseases result. Earlier in this chapter we described two diseases resulting from ecological disturbances. One was the mouse plague in South Australia which could have caused infections in humans. The other was the Black Death. Further ecological diseases will be described here and in other chapters.

Global warming

Global warming has grave consequences for the spread of malaria and many other diseases. Malaria is an infectious disease that has afflicted humans for thousands of years. In certain geographical areas of the world it is a serious infectious disease. But it is also an ecological disease because we acquire it from mosquitoes whose living conditions are sensitive to moisture and temperature. In some tropical countries clearance of rainforests and building of roads causes pot holes and craters in which water collects, thus providing breeding sites for mosquitoes. Malaria among humans has increased recently in these areas. But with global warming, it will spread to regions that are temperate at present. Its spread and enhancement will be caused by our disruption of ecological systems (see page 97). The mouse plague in Adelaide resulted from the ecological damage to pastoral lands. The Black Plague resulted from changing social and economic factors whilst the AIDS epidemic (see page 68) may have resulted from clearance of tropical forests and from globalisation.

A funny and possibly true story reported by Hunter and Amory Lovins, American energy efficiency experts, demonstrates the sometimes convoluted and unexpected relationships between various components of the Earth's ecosystem. They called this tale How Not to Parachute More Cats.

*In the early 1950s, the Dayak people in Borneo suf-
fered from malaria. Large amounts of DDT were
brought in and sprayed to kill the mosquitoes which
carried the malaria. The mosquitoes died, the malaria
declined, so far, so good. But there were side-effects.
Among the first was that the roofs of peoples houses
began to fall down on their heads. The DDT was
also killing a parasitic wasp which had previously
controlled thatch-eating caterpillars. Worse, the DDT-
poisoned insects were eaten by geckoes which were
eaten by cats. The cats started to die, the rats flourished, and the people were threatened by outbreaks
of sylvatic plague and typhus. To cope with these
problems, the World Health Organisation was obliged
to parachute live cats into Borneo.* [7]

The relationship between global hazards, human health and
ecological disease are extremely complex as is shown by the next
examples.

Acid rain

A global ecological disaster, unrelated to infectious disease, is
acid rain. Acid rain results from pollutants, oxides of sulphur
and nitrogen, which come from coal-fired and to a lesser extent
oil-fired power stations. They are dissolved in rain and now affect
the entire northern hemisphere, particularly temperate areas. The
pollutants that form the acid rain are carried on prevailing winds,
from the USA to Canada, from the UK to Scandinavia and from
Western to Eastern Europe. The economic growth of China is
increasingly producing acid rain which will affect its neighbours.

Within China it is causing widespread ecological changes.

When acid rain falls into streams and lakes they withstand the acid for many years because they contain buffers that counteract the acid. Eventually these buffers are exhausted, the water becomes very acid and the species living in it die. This is a hidden disaster. The streams and lakes remain clear and beautiful but are now devoid of fish and their ecological systems are probably irretrievably damaged. In turn this affects a whole range of species which depend upon these lakes and streams. The forests of the northern hemisphere are also sick, with an estimated one in four trees severely damaged – both by direct damage to foliage by the acid rain and by the leaching of trace elements from the soils. Arable land subjected to acid rain has shown a fall in food production over the past few decades. Acid rain also affects human health by leaching toxic metals into drinking water. In conclusion, the burning of fossil fuels has significantly changed ecosystems of the northern hemisphere and probably of China and will have an indirect and often unmeasurable effect on human activities for the foreseeable future.

Stratospheric depletion of ozone

The ozone layer is situated in the stratosphere above an altitude of 10,000 metres. It filters out and therefore protects the Earth from most of the solar ultraviolet radiation that is harmful to life. The ultraviolet radiation itself is responsible for the conversion of some of the oxygen in the upper layers of the atmosphere to ozone and the creation of the ozone layer. Chlorine containing chemicals, chlorofluorocarbons (CFCs) catalyse the breakdown of ozone. This means that each molecule of CFC can destroy in turn thousands of ozone molecules. CFCs are man-made chemicals used for refrigeration fluids, and propellants. Because ozone is slow to regenerate, the 'holes' in the ozone layer that have resulted from ozone depletion will be with us for many decades after the abolition of CFCs. The Montreal Protocol 1987 provided some degree of international agreement to reduce the emissions of CFCs by the year 2000; the protocol was revised in 1990 and 1992 with agreement on a phase-out of CFCs by developed countries by the mid-1990s.

Terry Bruce, a Sydney-sider aged 25, visited his family doctor to ask for vaccination against hepatitis. A machine operator, he was giving up his job having saved for three years to travel in Indonesia. He was single and had the wander-lust of many young Australians. Terry's doctor did the injections himself because the practice nurse was off sick for the day. When Terry's shirt was removed he noticed a round pinkish spot with some black colouring on the back of his left shoulder. 'You may have a spot there due to sunburn,' he informed Terry. 'To be safe it ought to come off. Don't worry about it, we'll be able to reassure you once it's done.' He made an appointment for Terry to see a colleague who was a plastic surgeon. The removal was simple and painless; the spot was cut off together with normal skin for two centimetres around it. Four days later, the surgeon's secretary phoned Terry asking him to return. The pathology showed malignant melanoma. Terry was reassured that all the tumour had been removed and asked to come back in two months and then every six months for five years, 'to make sure there is no recurrence'. Indonesia was postponed for a while. Three months later, a few days before setting off on his overseas adventure, Terry had a fit in the street and was admitted unconscious to hospital. A scan of his brain showed a single large tumour on the right side. A cell from the original melanoma had been carried by the blood stream to the brain where it had grown rapidly. After the fit he was left with weakness of his left arm which stopped him working again. An X-ray of his chest showed many other tumours because melanoma cells had also spread to the lungs. Terry died three months later. His exposure to sunlight seemed no greater than that of anyone else. He was not a regular visitor to the beach but had been burned badly on a few careless occasions.

The health effects on humans of increased ultraviolet irradiation are already apparent. Ultraviolet irradiation causes skin cancer. The US Environmental Protection Agency has estimated a three per cent increase in skin cancer and a one per cent increase in malignant melanoma for every one per cent decrease in the ozone layer. The ozone layer has shown major losses of ozone, for example 40 per cent over the Antarctic during parts of the year. People in both the southern and northern hemispheres are now affected. Exposure to ultraviolet irradiation also contributes to senile cataract which is responsible for 30 million cases of blindness worldwide. This disorder will increase greatly as ultraviolet irradiation increases. Finally there is concern for the adverse effects of increased ultraviolet irradiation on the human immune system, though as yet there is insufficient research on this topic. A suppression of immunity is likely to lead to an increased susceptibility to infection of the skin and ultimately to other infections.

People have the option of protecting themselves against increased ultraviolet irradiation by wearing clothing and suncream. Animals and plants do not. A majority of plants are sensitive to ultraviolet radiation and the yield of food crops will fall. The most profound and disastrous effect of ultraviolet radiation is likely to be occurring on the phytoplankton of the oceans. This will result in a chain of events that typifies the damage to all the world's ecosystems. Phytoplankton are the basis of all marine food chains; they take up a vast amount of carbon dioxide, acting as a reservoir or sink just like forests. It has been estimated that the reduced uptake of carbon dioxide due to a ten per cent loss of marine phytoplankton is equal to the annual carbon dioxide emissions caused by the use of fossil fuel. This will enhance the greenhouse effect. Phytoplankton release dimethylsulphides into the atmosphere which allow clouds to form, thus reducing ultraviolet irradiation. There is a chain reaction:

ultraviolet radiation reduces phytoplankton which produces fewer sulphides so fewer clouds are formed and the ultraviolet irradiation does more damage. The issue is complex and has wide implications, some of which may not have been predicted yet.

Ozone depletion is worthy of further discussion because it provides a model that is simpler than global warming and other ecological health problems. Looking at this problem also reveals the shortcomings in our thinking. We know the cause of the problem and how to resolve it but decisive action has proved difficult. Why? Although the cost of not acting promptly is large and may be immense, we are unable to act because of local economic consider-ations and the implications for governments and their needs for re-election. The power of balance sheets and jobs holds sway over the misery of hundreds of thousands who will die prematurely from melanoma or will be blinded by cataracts. The even greater problems of falling food production and of damage to the oceans are unseen and disregarded – they will burden others in future governments and generations. As we will see with global warming, those opposing reduction of CFCs most vehemently are in general the governments of developing countries which use cheap CFCs necessary for their 'development'. They want no impediment to joining the interna-tional 'growth' club. They ask, 'Why should we curtail our develop-ment when you have already developed?' Yet these countries will suffer most from ozone depletion. All these issues apply to the much greater problem of global warming (page 80).

The implications of ecological health and disease

Global cooperation and solutions will be required to deal with eco-logical health, but national sovereignty and short-term self-interest will prevent solutions. A generation of re-education may be required before the problems are recognised and solutions will depend on changes in our economic and value systems. Ministers of health and the environment are usually rendered impotent by the ministerial heavies from resources, finance, employment and by the leaders of industry. Ecological health should be the concern of all government departments. By contrast, public health problems, because they are simpler, are usually solvable within our present political and economic systems by departments of health and the environment.

However, there are many crucial links between the ecological health and population health. Ecological health can be regarded as on a continuum with population health. This is particularly so in relation to the many calamities inflicted on the human race. Famine is the most distressing image we see on television. The causes of famine are well recognised – over-population, poverty, economic systems that impose debt and cash cropping instead of adequate subsistence farming, inadequate yields of crops on degrading land leading to inability to store grain for the bad year, and many more. A vicious cycle is established whereby malnutrition, anaemia and infection result in the population suffering deteriorating biological function, physical inactivity and inability to work and think effectively. Public health can help this situation with the strategic provision of nutrition, sanitation and vaccination to prevent infection. However, famine itself is responsible for ecological damage and a vicious circle of damage to the environment results from the urgent requirement to provide food. The landscape appears devastated, and indeed it is. We pay little attention to the ecological consequences of wars, famines and social conflicts. Often the conditions essential for survival are destroyed.

Population health depends upon the discovery of new technologies, for example vaccines, and upon the implementation of planning, sanitation and social organisation. Ecological health will not be helped by these, indeed the acceptance of the idea of ecological health is prevented by a naive optimism about what humanity might be able to achieve by technology. There is every reason to doubt the ability of technology to solve our global problems. Technology has not solved fundamental problems of health such as enteric infectious disease, warfare or famine. Since health problems are environmental and social as well as biological, they will require social and environmental solutions.

The human follies that we predict will force us to change our attitudes are global warming, deforestation, ozone depletion, decline in biodiversity, and increased globalisation (travel, commerce) because they have the capacity to inflict ill-health upon all. Whether we have the capacity to predict these changes in detail — and have the social and organisational skills to respond to them — remains the great uncertainty of the last years of this millennium. We are perplexed that political leaders in advanced, wealthy countries such as Australia have not recognised the implications of these factors acting together. These factors will affect the stability of political, military and economic associations between countries. Failure to consider them will see history regard us as latterday Neros who fiddled while Rome burned.

The broadening of our idea of health — from the present emphasis on disease to a social model of the healthy human being — has to be expanded further because human beings are part of the ecological community. Not only are housing, nutrition, individual rights and democracy important for our health and well-being, but the nature of our relationship with the land and with its plants and animals is also important. This challenges what we are taught in our culture, which sees ecology as 'out there', and separate from us (see Chapter 7). But ecology influences the way we deal with each other and other life forms within the global ecosystem and its disruption causes grave health problems. Ecological health builds on, rather than discards, previous ideas of health. We recognise that good health includes our relationship with other living things on

Earth as well as how we organise ourselves socially. This has practical and ethical implications for all aspects of our lives. We begin to see that we have a debt to other groups of humans and to environments that needs acknowledgement. We begin to see that we need sensible, useful and sensitive rules about how to live. Environmental and health problems therefore cannot be separated from social, political and cultural developments and situations.

1. Ian Kennedy, *The Unmasking of Medicine* (Allen & Unwin, London, 1981).

2. H. Mahler. Keynote address to Second International Conference on Health Promotion, Adelaide, 1988.

3. Mason Durie. *Whaiora, Maori Health Development* (Oxford University Press, Auckland, 1994).

5. Basil S. Hetzel, 'Iodine Deficiency: a Global Problem', *Medical Journal of Australia* 165 (1996), pp. 28–29.

4. A.J. McMichael, *Planetary Overload* (Cambridge University Press, Cambridge, 1993).

6. L.H. Lovins and A.B. Lovins, 'How Not To Parachute More Cats: The Hidden Link Between Energy and Security, in *Toward a Postmodern Presidency: Vision for a Planet in Crisis* (University of California, Santa Barbara, 1989).

CHAPTER 2

WHAT IS ECOLOGY?

Ecology is the scientific study of the ways in which all living things interact between themselves and with their environment. The relationships between living things is extremely complicated. Imagine a patch of tropical rainforest. Within it there are thousands and thousands of different forms of life in the trees – insects, birds and animals. There are many different types of trees and plants. There are billions of worms and insects, fungi and bacteria. All depend upon their environment of air, water, the nutrition in the soil and energy from sunlight. This situation has been described as 'the web of life'. A spider's web has strength and cohesion and every strand is, in some way, linked to all others. In the web of life the relationship between different living things (species) is also carefully balanced. However, unlike the spider's web, the web of life is changing slowly all the time: some species are increasing whilst others are in decline.

The expression 'balance of nature' sums up ecological relationships. Within the web of life, one species is dependent upon others for food, and the web provides a means of survival for all. The words 'food chain' are often used to describe the process whereby soil, which contains billions of bacteria and fungi to maintain its structure, provides nutrition for many species of plants which are then eaten by many species of animals. At the top of this food chain are the predators, which eat other animals. Our children may obtain their first understanding of the food chain from the rhyme 'There was an old lady who swallowed a fly . . .' – followed by the spider to catch the fly, the bird to catch the spider, the cat to catch the bird, and so on. The natural food chain is usually stable because it contains a huge variety of species with many sources of food. But it may be unstable or fragile because many species are interdependent; if one key species declines, many others will also decline.

An apparently simple ecological system involving plants, insects, animals and birds in a particular environment such as a rainforest is very complex and difficult to describe. We therefore begin with a simple form of life: bacteria, the organisms that are everywhere in soil, water, air and in the human body.

Dora Black developed rheumatoid arthritis when she was 35. It began with stiffness in her fingers which soon became painful and swollen. Pain then developed in her shoulders and knees. By the age of 55, despite all the treatments medical science could offer, many of her joints had been destroyed by the ravages of the arthritis. She could walk a few yards with the aid of a stick but her husband had to help her shower and dress. She could no longer work, drive or help significantly with household chores. One day, she started to experience a burning sensation when she urinated and she had to go to the toilet much more frequently than usual. Her family doctor prescribed antibiotics but after a few days she developed diarrhoea. This rapidly became worse so that every one to two hours she had to struggle to the toilet. Soon she was passing blood in the faeces and she was admitted to hospital. Her belly was swollen, she was nauseous and had a high temperature. X-rays showed that her colon was greatly distended with gas and was about to perforate. To save her life emergency surgery to remove her colon was necessary. She was left with an ileostomy – the intestine discharged to the front of the belly and she wore a bag to collect the faeces. Because of her arthritis, the bag had to be changed twice daily by her husband. Dora had developed a form of colitis due to the antibiotic disturbing the balance of normal bacteria in her colon.

Faeces and the human colon

Faeces are stored in the colon and in most people are passed daily. They contain the waste products of what we have eaten. In fact faeces consist mainly of bacteria and the colon of each of us contains millions and millions of bacteria. Indeed, there are more bacteria than cells in our bodies. We are outnumbered! Human faeces illustrate a most intricate ecological system within the human colon.

A proportion of our food, particularly roughage such as wheat bran, cannot be digested and absorbed easily by the small bowel, so it passes into the colon where it is eagerly used by bacteria. The bacteria multiply until all food passing into the colon has been utilised and so the volume of faeces increases. There are so many species of bacteria and other organisms present in the colon that we cannot count them accurately. All these species live in harmony with their host and themselves by controlling each other. Some bacteria produce chemical substances which control other bacteria. The human body produces antibodies and other chemicals which slow the growth of the bacteria and prevent them from invading the body. The bacteria in the colon are an example of a stable ecological system that defends itself against invaders. It is very difficult for a new organism to establish itself in the colon and when it does, it is often an organism that causes disease. Diarrhoea or even dysentery (the passage of fluid faeces and blood) occurs.

Each of us inherits the organisms in our colon from our mother. They exist as a colonic 'blueprint' that stays throughout life and cannot be changed unless a powerful destroyer is used. After use of laxatives or a vigorous washout of the colon with an enema, the same bacteria return to fill the colon – they multiply from the remaining organisms residing within the folds of the lining of the colon and the 'blueprint' is used to restore harmony.

However, antibiotics will destroy many of these defensive

organisms; on occasions so many are destroyed that the ecological system in the colon becomes unstable, allowing one or more destructive organisms to enter and multiply rapidly. It is like using a pesticide on acres of crops – most of the native insects are killed, putting the crop at risk from a newly invading insect. The pesticide has created room for the invasion. In the human colon broad spectrum antibiotics, which destroy many of the organisms usually present in the colon, allow the multiplication of damaging organisms. One such organism, *Clostridium difficile*, causes a bloody colitis (inflammation of the colon) in a small proportion of patients who take antibiotics and diarrhoea in a much larger proportion. The normal bacteria which prevent invasion by pathogenic bacteria have been reduced and disease results as in the case of Dora Black. The balance of nature or the ecosystem of the human colon has been disrupted.

Just as in the human colon, intricate balances exist between thousands of living species in forests, pastures, lakes, rivers and the sea. The web of life is strong but once disrupted the results are unpredictable and can be dangerous. Disturbance of an ecological system leads to invasion by other species. And the most obvious circumstances are the outbreaks of so-called pests on cultivated land when the habitat of communities of animals and plants has been greatly simplified by man.

A second and much more important example of disturbed ecology in humans also arises from the use of antibiotics over recent decades; this disturbance threatens the future successful use of antibiotics.

Victor Brewster was riding his motor cycle to work when he rounded a corner and crashed head-on into a car overtaking a van and on the wrong side of the road. He landed 20 metres down the road with fractures to the bones of his face, pelvis and right thigh. He also had a tension pneumothorax caused by fractured ribs sticking into his lung. In this condition, pressure builds up in the chest and soon leads to a failure to breathe. Fortunately the accident

occurred near a large public hospital and Victor was soon in the
intensive care unit. An opening was made into his wind-pipe so he
could breathe, a tube was put into his chest to relieve the tension
pneumothorax and a catheter was put into his bladder. Victor was a
fit man, aged 32, and his chances of recovery were now greater than
90 per cent. He would have surgery for his fractured face and leg
over the next few days. However, he developed an infection in his
chest and antibiotics were prescribed. He failed to respond despite
seven changes of antibiotic. He had a high fever, pneumonia and
then blood poisoning (septicaemia). He developed an abscess in
the lung and one in the kidney. His blood pressure fell, his heart
failed and he died. A germ called Vancomycin resistant entero-
coccus (VRE) had gained access to his bladder via the catheter, and
spread into his blood and throughout his body. All known antibiotics
are totally useless against VRE. VRE now resides in many large
hospitals world-wide.

Bacteria that are resistant to many antibiotics are being
increasingly detected in our hospitals. This is related to the persistent
and widespread use of antibiotics. Some bacteria undergo genetic
changes, which allow them to resist antibiotics, and then multiply
rapidly. Such genetic changes will always be much more rapid than
the ability of the pharmaceutical industry to discover and synthesise
new antibiotics.

We expose bacteria to antibiotics not just in our hospitals,
but in our sewerage systems and in our ground water because of their
use in animal husbandry. Antibiotics similar to those used in humans
are widely used in animals to promote growth and a better yield of
meat in Western countries. The overuse of antibiotics in clinical
practice and in animal husbandry can lead to the emergence of resistant germs. Resistant germs in animals can be transferred to humans
in meat and milk. Germs that have developed a resistance to most
antibiotics can transfer this resistance to other germs. This has led
to the opinion that in future we are likely to be forced to return to

the surgeon's knife to treat many infections. The diseases caused by antibiotic resistant bacteria can be thought of as ecological for they thrive when the normal inter-relationships between organisms are changed by antibiotics in the environment, in animals and in humans. It is instructive to reflect on medical practice in the days before antibiotics.

Joseph Lister, born in 1827, practised in Scotland and was one of the great figures of surgery. In fact, it is often said that the entire history of surgery can be divided into two epochs, before Lister and after Lister. In the epoch before Lister, a majority of people undergoing surgery died of infection in the following days or weeks. A few days after the operation the wound would become painful and inflamed and then blood poisoning would ensue. In those days an infected cut on the hand could be a death sentence. As infection spread, the limb might be removed to save the patient. Abscesses were 'laid open' – that is, the infected part was cut open and left open to allow the pus to drain out. Patients lay in wards for weeks with open wounds from which drained stinking pus. In those days a compound fracture – when the end of a broken bone sticks through the skin – was almost a certain death sentence from infection. Lister introduced sterilisation so that the surgeon's hands, the instruments and the patient's skin were clean. He used carbolic acid.

James Greenlees, aged eleven, came into the Glasgow Hospital with a compound fracture of the lower part of his leg; he had been run over by a cart. The end of James' broken bone had come out through his skin leaving a hole to the surface. Lister put carbolic acid into this hole and onto the surrounding skin, applied a dressing and covered it to prevent evaporation. James recovered without any complications. The date was 12 August 1865. This was the first time carbolic acid was used and the post-Lister era had begun.

However, there were many infections for which carbolic acid could not be used. With infection of the lungs, pneumonia, a life and death battle raged for many days between the patient's immune system and the infection. Eventually, if the patient was to win, a crisis would occur with high fever followed by a sudden fall in temperature. Otherwise fever would continue and death would follow. Infection of the brain (meningitis) or of the abdomen (for example peritonitis) often meant death. Infection after childbirth (puerperal fever) was common and many young women died. The successful treatment of all these conditions was to await the antibiotic era nearly 100 years later. Many experts in infectious diseases now believe that we may return to the period between 1865 and 1945 for we will have 'sterilisation' but no antibiotics.

Ecological science and its limitations

Ecology describes the existence and behaviour of species in response to their environment and food supply; it describes their growth in numbers and their interactions with each other and with other species. In ecology there is a sense of 'everything connected to everything'. A member of a species lives in a population of that species which is part of a community of other species. Individuals and species interact, sometimes benefiting, sometimes harming other individuals and species. Humans are part of this interaction. The food resources and environmental conditions influence the health and well-being of individuals and species and hence their ability to reproduce. Competition for food and other resources for the most part limits domination by any one individual or population.

Because ecology is so complicated, scientific ecology is limited in the information it can provide for us. While ecological science, like most other areas of science, has progressed, it has come up against the unpredictability that arises when a vast number of variables and processes interact. Science or any intellectual pursuit can never take account of all these things at any one time. This situation leaves us with a dilemma about how far to go in seeking to understand ecological systems. The slow search for such knowledge hinders or delays the progress of measures which will stop ecological damage because governments wait for scientific proof. Yet because preservation of ecological systems is now vital, we must have political and social solutions.

Other views of ecology

As we will see in later chapters there are alternative, non-scientific views of ecology. There is a popular view, a romantic view of nature. Charles Darwin's description of life on the bank of a river is this type of description because it is so colourful as well as scientific.

> It is interesting to contemplate an entangled bank, clothed with many plants of many kinds, with birds singing on the bushes, with various insects flitting about, and with worms crawling through the damp earth, and to reflect that elaborately constructed forms, so different from each other, and dependent on each other in so complex a manner, have all been produced by laws acting around us.[1]

This popular viewpoint is very valuable because it helps us appreciate the natural world for its own sake rather than as a resource we can use simply for our own needs. The word 'ecology' is also used by individuals and groups who seek to change the political and economic landscape and individual behaviour by describing 'ecological' politics and ethics that are different from our current forms of political, social and economic organisation and individual behaviour. They have this focus because they are concerned about the way in which our Western society treats nature and disturbs the ecological system of which they are part.

In contrast, those who promote 'deep ecology' see nature as in some way having a value akin to the value we place on human life. That is, the rivers, the rocks and other components of the Earth have some right to existence. If we think in this way, we will sustain the ecological systems because humans have a 'proper' place in the scheme of things – and that place is definitely not at the pinnacle of biological evolution! (We will discuss these examples only when they further our goal of developing ideas on ecology and health. In the main, such ideas do not relate to health and this is one of their limitations.)

One task of this book is to outline an ecological way of living for this is essential to health and well-being. It is not practicable for some of us to live in an ecologically sustainable way while a vast number of people live and die in abject poverty, misery and suffering. At the same time we believe there is enough evidence to show that we do not have to destroy more of this planet's ecosystems to adequately look after all the humans that inhabit it. As the world's population grows the task of finding solutions to these extraordinarily difficult problems will become even harder. This is because the increasing number of individuals and their demands increase the complexity of the problem. There is no way to arrest population growth suddenly other than by the decrees of dictators, so for the moment the task is to adapt humans to care for the environment and ecology in a more populous world. The task is formidable for there may be little time left to accomplish it.

A case of mad-government disease

The complexity of ecological diseases and the difficulty of predicting them is well illustrated by the following.

Thirty-four-year-old Janice Sims, who lived in Glasgow with her two children, had an outgoing personality and many friends. When she became anxious and irritable her family doctor thought she was suffering from depression and treated her with anti-depressant drugs. She became worse, she lost memory, heard voices that were not there and had a feeling of 'pins and needles' in her hands and face. She went to live with a friend who looked after her but was soon admitted to a psychiatric hospital. Tests, including a brain scan, failed to result in a diagnosis. Her condition worsened, her memory failed totally, her breathing was affected and she died 11 months after she first went to her family doctor. A post-mortem showed that her brain was affected by Creutzfeldt-Jakob disease, the human form of Mad Cow Disease.

Mad Cow Disease arises from interference with the food chain. It is caused by a prion. Prions are remarkable small proteins which are responsible for several diseases of the brain and which are transmitted from animal to animal and between species of animals. Prions do not carry any genetic material and are therefore different from bacteria and viruses. Mad Cow Disease, Kuru and Creutzfeldt-Jakob disease are all caused by prions that infect the brains of humans or animals. They cause many holes in the nervous tissue and the brain looks like a sponge. Hence the medical name of Mad Cow Disease – bovine spongiform encephalopathy (BSE). Kuru was a disease found in the highlands of Papua New Guinea caused by cannibalism. People who ate the brains of other people infected by Kuru would develop Kuru themselves. In Kuru there is a rapid loss of coordination and the individual loses balance, cannot walk and also develops dementia. Creutzfeldt-Jakob disease is similarly characterised by dementia and loss of coordination, until now it was thought to be genetic with perhaps one case occurring in every million humans. Scrapie in sheep and goats is another prion disease. The animal loses coordination and becomes incapacitated.

In the UK in the 1970s, the meat industry, unknown to the consumer, began feeding protein supplements to cows. These supplements were obtained from rendering the carcasses of sheep, some of which had died from scrapie. By the mid-1980s, 130,000 cattle had been afflicted with a disease similar to scrapie which was dubbed Mad Cow Disease. It led to the cows going 'mad' – falling over, and being uncoordinated and highly irritable. In 1988 an official report recommended that the carcasses of the affected animals should be destroyed and that there should be a ban on feeding animals with protein derived from carcasses. The authors understand that these measures were not carried out by a governmental system that was unduly influenced by market forces. 'Business' and exports would be adversely affected. Members of the government knew the worrying facts yet refuted them. One senior minister had the TV cameras record his young daughter eating a beef burger to reassure the nation. Not only the cows were mad!

Two recent events have raised the alarm. It has now been shown that the scrapie prion, which is presumably causing Mad Cow Disease, can be transmitted to a variety of other animals. Secondly, Creutzfeldt-Jakob disease has for the first time been found in young persons, such as Janice Sims, in the United Kingdom. In the past, cases of Creutzfeldt-Jakob disease have only occurred in middle-aged and elderly persons. Looking at the available evidence it seems likely that this prion has been transmitted from beef to humans and will cause an increasing number of cases of Creutzfeldt-Jakob disease in young people. The worst-case scenario is that there could be thousands of such cases.

Could these events have been predicted? In fact many warnings were issued by scientists in the 1980s. From the experience of Kuru in Papua New Guinea it was already known that infection could be transmitted from human to human by eating brains, so why not from sheep to cow to human? Mad Cow Disease is an ecological disease. It is unnatural to feed a cow, which is a strict herbivore (eats only grass) with animal protein. This is imposing upon the cow the potential for infections which it has not experienced before and to which it can have no resistance. The demand for high-tech means of increasing food production will increase as world population increases. These new methods require careful examination and control if they are to be safe. When in doubt, the precautionary principle should always be adopted. Yet our free-market system has cut back on regulatory systems worldwide in favour of allowing the market to manage itself.

Ecological sustainability – what is it?

We refer frequently to 'ecological sustainability'. Concern about sustainability can be summed up as: what will happen if we continue to live in this way? Will we run out of resources? The world's population has the possibility of growing exponentially but the resources to sustain this increase are finite. If we were any other species, our growth in population would have been controlled by lack of food and resources but our ingenuity to develop science has allowed us to continue.

The Annual Review of Ecology and Systematics described ecological sustainability as: living within the limitations of the natural environment; maintaining the natural environment to provide resources and to act as a sink for wastes; harvesting renewables, for example forests, within the rates of regeneration; harvesting non-renewables (e.g. a particular metal or form of energy) at rates no greater than the rate at which substitutes can be found or created.[2]

We should ask whether individual, collective, local, regional and global activities are ecologically sustainable.

Ecological science helps us to answer questions about the sustainability of agriculture, fisheries and forestry. The Ecological Society of America has formulated a set of questions which it hopes will take us further along the path to sustainability. Called the 'Sustainable Biosphere Initiative', its main focus is on the necessary role of ecological science in the 'wise management of Earth's resources and the maintenance of Earth's life support systems'. It calls for basic research towards the acquisition of ecological knowledge, communication of that knowledge to citizens, and incorporation of that knowledge into policy and management decisions.

Again we see that the initiative will not succeed without the understanding and cooperation of politics and government.

Gaia and the global ecological system

An influential approach to ecological sustainability has been to think of the planet as one ecological system. The most well-known of these approaches has been the Gaia idea (named after the Greek goddess 'the Earth'), proposed in 1988 by James Lovelock, an English scientist. In his words:

> Gaia is the Earth seen as a single physiological system, an entity that is alive at least to the extent that, like other living organisms, its chemistry and temperature are self-regulated at a state favourable for life. [3]

Lovelock thinks of the entire Earth as one single complicated living thing. We could think of the human colon as a single living thing which contains a complex ecological system with each bacterium functioning to regulate others and keep them under control. Similarly, each component of the Earth functions to maintain stability by interacting with all others. Each part of the Earth depends on all other parts – the oceans, forests, atmosphere, soils, etc – and this interdependence embraces all species of life on Earth including us, the humans. Self-regulation of the Earth is automatic in the same way that our human body controls its temperature, breathing, eating and excretion. When the temperature rises in the human body, blood vessels at the surface of the body are opened to transfer heat to the surface; sweating occurs and evaporation of the sweat cools the body. When body temperature falls, blood is moved away from the surface and our muscles shiver to create heat. All these events happen automatically and spontaneously and are orchestrated by a thermostat in the brain so that our temperature is kept constant. The Earth, too, has mechanisms to maintain the temperature of the atmosphere within narrow limits. When global warming occurs over millions of years, due to a rise in the concentration of carbon dioxide (see Chapter 4), rainforests slowly spread north and southwards, trapping the carbon dioxide in their increased growth; global warming is controlled naturally by a series of interrelated mechanisms.

There has been considerable discussion and disagreement about the Gaia hypothesis. It is easy to ridicule it by asking how can the Earth be a single living thing. We are sure James Lovelock intended us to think of it that way in order to make us think of the planetary ecosystem. He intended us to think of Earth as a single web of life, one that is strong and complex but which, like any other ecological system, could be susceptible to substantial damage.

Ecologists have observed that ecological systems, when placed under various stresses for long periods, show a decline in the number of species although the system continues to function. In some instances, decline proceeds to a massive loss in the number of species. An analogy in human medicine would be the human liver,

which we can subject to the most amazing barrage of destruction by
alcohol. This steady destruction is accompanied by the remaining
liver doing all that is necessary to purify the blood and keep us alive
until perhaps 80–90 per cent is destroyed. Then, the consequences
are suddenly revealed: many, many biochemical disturbances
occur and the patient soon dies. Only a liver transplant can save
this person. An ecological system, and indeed the Earth itself, has
these reserves and can sustain its functions until a specific point of
danger which could arise now, in 50 years or 100 years. Who is to
say? Instead of a liver transplant, humans may need to transplant
themselves onto another habitable planet.

Ecology, Gaia and modern science

Generally, the public believes the results of science, yet it should be
remembered that once upon a time scientists pronounced the Earth
to be flat! Science itself undergoes constant revision as new knowledge emerges. The Gaia hypothesis casts doubt on some of the
assumptions and processes of modern natural science. Good heavens,
you may say, to question science questions our understanding of
modern life. It is a call for revolution! Yes, that is just what we are
proposing.

One has to understand the limitations of the scientific
approach to environmental problems. A stable ecological system
may have thousands of living things interacting in the web of life.
The common scientific approach is to study the inter-relationships of
a handful of these living things. A hypothesis is proposed as to how
A affects B, C and D and simple studies are carried out to prove or
disprove it. The emphasis is on *simple* studies – for if there are too
many factors (variables) at work it will not be possible to interpret
the results and obtain definite answers. In practical terms, the

scientist will be accused of poor design of the experiment, the results will be rejected by the scientific journal, his/her standing will fall and if this poor performance is repeated many times, there will be no promotion! Thus science studies the immense complexity of the web of life by studying separately each of tens of thousands of pieces of a jigsaw puzzle; when the puzzle is completed, then we will understand the web of life – or so it is thought. When applied to science in general this has been called the reductionist approach for it reduces the complex to its simple, individual parts.

An ecological approach, on the other hand, recognises that a living community or an ecological system is much more than the sum of its parts. Therefore it must be considered as a whole, but it is very difficult to develop scientific methods to do this. In the study of the human body, for instance, science has described a large proportion of the components, the cells and the chemistry, but this tells us little about the integration of all these parts and how we function as whole, individual humans. In terms of the Earth and ecological disease, we need to think of the planet as a whole and solutions to problems will be global ones.

One of the most expensive scientific experiments ever conducted, Biosphere 2, has important lessons for our understanding of ecology. Biosphere 2 was designed as a three-acre airtight system which cost 200 million dollars to construct. Enclosed within it were ecological systems, comprising plants, animals, insects, water, atmosphere and eight humans; the ecological system was intended to function independently from the natural world and Biosphere 2 was expected to be self-sufficient. The project would allow an entire ecological system to be studied, thus avoiding the problems associated with studying parts of an ecological system. As time passed, life in Biosphere 2 became unstable. Some plants became aggressive and overran others. Most vertebrates and insects became extinct as did many birds. Crazy ants and cockroaches took over. The ecosystem as we know it had collapsed. The humans had a greatly increasing work load to produce enough food and pure water to survive. The conclusions are obvious. An ecological system in isolation is not sustainable, or rather we do not have the scientific understanding to sustain it. To retain our life-support systems, there is no alternative but to maintain the viability of the Earth.

There are two further limitations to the scientific approach to a complex problem. Firstly, it may produce conflicting results and those with vested interests will call for caution and delay till further studies are done. Secondly, we have to recognise that the 'independent' scientist today is in a minority. The majority are employed by industry, multinationals and government departments with their own agendas – such as 'we must continue this industrial process but go and study its effects on wildlife'. How many scientists will have enough integrity to come back to the company with preliminary results saying, 'Stop your process, I have some evidence that it's causing a problem?' In any case, there are so many other important factors to salve consciences with – we need to create jobs, there is no alternative product, the balance of payments is poor. At the extremes we have scientists advising organisations such as the Forest Protection Society in Australia, which is an organisation supporting felling and wood-pulping of our old growth forests, and which maintains that clearfelling trees is good for forests.

In fact, the independent, 'free-thinking' scientist is increasingly rare in our universities. Molecular biologists, for instance, increasingly have financial stakes in the biotechnology business and so there are conflicts of interest between university and public policy.[4] Indeed, it might be asked whether this conflict of interest extends to the reporting of scientific results. We fear that the 'successful' universities of the future will be efficient, entrepreneurial companies which manipulate information for gain as successfully as large private-sector companies. Increasingly, education can be seen as serving greed, growth and the enhancement of capitalism (we discuss this further in Chapter 7).

Loss of species – is there an ecological crisis?

Some scientists such as Norman Myers believe that there is evidence that 20–25 per cent of all animal species on the Earth will become extinct in the next 30 years, and 50 per cent by the close of the twenty-first century. These estimates are based on extrapolations from data of present day losses of habitats. Of the estimated 10 million or more species existing on the Earth at the moment, approximately half live in tropical forests, which in the 1980s were being destroyed at the rate of about one per cent a year worldwide but at much higher rates in some areas (for example, three per cent in Nigeria and India). There is evidence that when the remaining stands of forest have decreased to a certain size, loss of species will accelerate. This is because of reduced habitat, less protection, the ease of encroachment of other species and the inability of species to reach and breed with the same species beyond their own local areas. This is why conservationists are so keen to maintain wildlife corridors linking areas of native vegetation within farming areas and cities. The rate at which species are lost will further accelerate because the loss of each species will then lead to loss of other species dependent on it. Some scientists refer to the rate of loss of species as a 'mass extinction'. We have to recognise that once a species is gone, it is gone for good.

With the loss of many species, each of which is necessary for the successful functioning of others, including humans, a point may be reached at which the sustainability of the Earth's ecosystem as a whole is impaired to the point of malfunction, resulting in human ill-health on a large scale. There will be many ecological diseases as yet unthought of.

These considerations raise the issue of human responsibility. There are two alternatives: we can either be part of the web of life or we can sit above it and try to control it. We have escaped the constraints which keep us under control by ecological processes. If we are a species like any other, the population explosion would be followed by a period of contraction to levels which could be fed a reduced production of food. But we have postponed this contraction by technological advance and by the use of irreplaceable resources such as soil. Therefore, we are sitting above the web of life and have placed ourselves in the position of needing to be clever enough to control the consequences of loss of soils, species and all other calamities. We are no longer in a web of safety. The alternative is to acknowledge that we are part of the web of life and accept responsibility for controlling population growth and energy use.

Thinkers and philosophers have two ways of looking at our increasing loss of species. The 'sit above' theory believes that money makes the world go round. So species loss has been estimated in economic terms. How many drugs of benefit to us will be lost? (about half our pharmaceuticals are derived from plants, bacteria, fungi and other species – see Chapter 3). How many industrial chemicals will we lose? How much breeding stock for grains and other foods will be lost? This may be how the issue is seen by the hard-nosed economist, the destructive multinational company or the ignorant politician. Many others will believe the situation can be retrieved by genetic storage or by genetic engineering.

The other way of looking at the issue is to see that humans are responsible for retaining species because we recognise our partnership with the global ecosystem. We believe that our presence on this Earth depends upon our relationship with other living things, respecting them and recognising the interdependence of all life on Earth. This discussion is expanded in Chapter 3. There is a relationship between the global ecosystem and our health and well-being on this planet. At the extreme, calamity may befall much of the human population with sudden and excessive global warming caused by our continued burning of fossil fuels, our destruction of rainforests, the degradation of soils and water and the like. Even short of calamity, there are processes occurring on Earth that will affect our health and well-being. We discuss them in subsequent chapters.

In *The Sixth Extinction*, Richard Leakey analyses the world-wide episodes of extinction that occurred millions of years ago.[5] The age of the dinosaurs probably ended when an asteroid hit the Earth and the resulting cloud of dust encircled the Earth, producing a prolonged winter. Other periods of extinction probably resulted from climate change – a warning to us all. However, the most recent mass extinction (apart from that which we are causing now) probably commenced about 11,000 years ago with the spread of the human population worldwide. Hunting led to the extinction of many species of animal, particularly in the Americas but also in Asia, Australasia and northern Europe.

The arrival of the Maoris in New Zealand was followed by the hunting to extinction of many species. How can we reconcile this with the Maori adapted to the environment as described in Chapter 1? Essentially the spreading human race consumed all before it. The various species of animal were easy to kill for they had not experienced human predators. After thousands of years, these indigenous peoples then adapted to their environment. Today the expansion and activity of 'Western' man is creating another mass extinction of species, but on a much greater scale.

Cities and ecology

Previous civilisations organised themselves in cities which required significant environmental resources to sustain them – water, forests and land for food production. Mediterranean civilisations like that of Rome cut down and utilised their prolific forests and then eroded the soils by cultivation or grazing. Eventually the urban civilisation could not provide for itself. In other civilisations in the Middle East water probably became a limiting factor and the increasing need for it would have led to conflict between communities and nations. These past civilisations were separate from each other in time and location and no overwhelming harm came to the Earth as they rose and fell. Today there is one global civilisation and we will rise or fall together for the harm we do is global. We can see that the changes in the Earth that threaten us all – global warming, loss of the ozone layer and pollution – arise primarily from urban civilisation and its demands for energy and transport. Indeed, three-quarters of the world's pollution comes from urban society.

Many writers and physicians now recognise that humans are part of a closed ecological system on the Earth. In *Public Health and Human Ecology*, Last described the situation as follows:

> *Ecology is concerned with the healthy interaction of creatures in a closed system. Human ecology includes humans in this system. Humans interact with each other as well as with other living creatures and these interactions have important effects on all partners in the complex closed ecosystem of our planet. We ignore this reality at our peril.*[6]

We must think of human society as an ecosystem, with each individual connected to and dependent upon many others. About a half of the world's population lives in cities and it is easy to see that these cities function in numerous ways that negate all the principles of ecological systems described in this chapter. The power stations of urban civilisation produce a huge amount of energy which allows the cities to function but also leads to global warming. Urban sprawl, which leads to use of automobiles, also contributes greatly to global warming. Road and concrete surfaces increase run-off of rainwater to cause pollution of rivers, lakes and seas. Water resources are overused by pumping of ground water, and storage of water by dams destroys river valleys and thus ecology. Costly sewerage schemes are needed to prevent pollution of coastal areas with loss of coastal habitat. Landfills and dumps lead to contamination of soils and loss of amenities. Urban sprawl eats away at good agricultural land. There is a chain of events whereby our urban living leads to global warming and pollution. The demands of our urban civilisation for fine woods can be linked to the loss of our forests.

For all these reasons the minds of some are turning to 'urban ecology', that is to the methods by which we can live sustainably without causing such damage. Urban living will have to utilise solar energy, conservation of energy and water and the use of grey water, biodegradable materials, local production of foods from solar greenhouses with composting, locally managed sewerage recycling schemes and local production of goods. These necessary changes will challenge today's economic system for the new concept will be one of minimal utilisation of resources and of recycling involving water, wastes and materials. Urban development will be community oriented and ecologically sustainable, instead of being just an arm of the profit system. Like the health of our farmlands and our seas, the change in development and maintenance of our cities and urban lifestyle will depend upon a reform of our collective philosophy and economic systems (see Chapter 6).

Psychology of cities and urban living

In many ways, city living is a complete opposite to ecological living. People are crowded together with no immediate links between them. Even living close together in blocks of flats, individuals remain isolated. It is anonymous living. Such living often reflects the competitiveness of society with everyone out for themselves. All those attributes fostered by our society, and in particular, each individual's wish to succeed and be top of the pile, is developed and fostered. Cities have a life cycle of creation and destruction. They expand and so need more cars and roads. The more established communities with a modicum of social interaction are disrupted by new highways and developments demanded by business. The quality of life deteriorates and those who can afford to, move to the suburbs. The big city becomes a place of isolation, poverty, crime, thoughtlessness and despair. The psychological effect of cities has not been studied from an ecological perspective. The attachment to the land and a feeling for the land that most of us retain to some extent and which is so evident in indigenous people is perhaps reduced or lost in city dwellers, although they may have a sense of belonging to 'place'.

It is interesting to look at the behavioural patterns of many of the inhabitants of the centres of cities. Tiny window boxes provide a touch of green. A straggling hotchpotch of roof gardens is evident on many tall buildings as one flies over a large city. These are not being cultivated to provide the poor with vegetables, they are providing the human spirit with something growing and alive apart from the seething mass of humanity. The psychology of the colour 'green' also warrants further study. There is some evidence that it induces a sense of calm and serenity.

Pet ownership also burgeons in such cities, not just the inevitable dog or cat but the whole range of exotic animals are hidden in the labyrinth of flats and high-rise buildings. Are these the instinctive remnants of days when we fitted better into the ecological web and had permanent relationships with the animal kingdom? There is something instinctual and beneficial in this relationship as seen by the studies that show the lowering of human blood pressure conferred by pet ownership, and indeed people who have such relationships appear to live a more satisfactory and longer life.

The cities are the battleground to sustain the environment as much as the forests, lakes or shorelines. To date nearly every large city has gone the same way. It might pass through a phase where living standards are reasonable in terms of environmental conditions but the free market and political expediency soon wrecks those pleasant places. If we look at Adelaide, one of the best planned cities in the world and perhaps one of the most acceptable environmentally, a central area of a square mile is surrounded by a ring of parklands a mile wide. It was planned that way in 1836. Today these parklands are under constant threat of erosion and attrition at the edges. In times of financial hardship any plot of greenspace that can be sold is rapidly swallowed up by the developers.

Cities become uninhabitable and unhealthy because of pollution and transport and because their open spaces disappear. People have nowhere locally to go for simple exercise and recreation and they have to travel by motor transport to their gyms or to the periphery of the city. Cars produce particulate pollution as well as ozone and both are damaging to human lungs, conferring ill-health on the many poorer people who have to remain in central areas of cities. Wide streets, green trees and grass all act as a sink for these pollutions and provide a healthier city. Yet where does this appear in urban planning and thinking? It is peripheral. There is always a good reason to chop down a tree – the road is not wide enough, a branch might fall on someone, it obscures someone's view or its roots are impeding the underground pipes.

Now these thoughts may seem to be very disparate from the discussion of ecology at the beginning of this chapter. However, in global terms all these problems are intertwined. The problems of the cities flow into our forests and our atmosphere and unless resolved will lead to the destruction of the ecosystems of the world. They are, therefore, a major factor in the production of ill-health.

---

1. Charles Darwin, *The Origin of Species by Means of Natural Selection*, 1859.

2. *The Annual Review of Ecology and Systematics* (Annual Reviews Inc, Palo Alto, California, 1995).

3. James Lovelock, *Gaia: A New Look at Life on Earth* (Oxford University Press, Oxford, 1979).

4. R.C. Lewontin, *The Doctrine of DNA: Biology as Ideology* (Penguin, London, 1993).

5. R. Leakey and R. Lewin, *The Sixth Extinction* (Weidenfeld & Nicholson, London, 1996).

6. J.M. Last, *Public Health and Human Ecology* (Appleton & Lange, East Norwalk, Connecticut, 1987).

CHAPTER 3

THE LUNGS OF THE EARTH: BIODIVERSITY, VEGETATION AND FORESTS

'Save the bush' and 'save the forests' are recognised by environmentalists as more than important environmental messages – they are health messages. In this chapter we will trace the relationship between our forests and other vegetation, and our health and well-being.

Forests and native vegetation

Why so much fuss about forests? Because they are both a reservoir of biodiversity and an important contributor to the stability of the Earth's atmosphere – and as a result to our survival. Therefore the struggle to save them is intense. Their preservation forms the pinnacle of the aspirations of the conservation movement because it is seen as vital and more important than political or ideological struggles such as communism versus capitalism, both of which have been destructive to the environment.

Destruction of the forests is brought about by people who are ignorant about the effect of their actions, by the poor who feel they have no alternative, by the greedy who will plunder anything, and by governments who wish to balance ill-conceived budgets.

We will return to the economic aspects of these points later. The destruction of our forests also relates to human attitudes. The American Henry David Thoreau, a forefather of the conservation movement said:

> *If a man walk in the woods for love of them half of each day he is in danger of being regarded as a loafer, but if he spends his whole days as a speculator, shearing off those woods and making the Earth bald before her time, he is esteemed an industrious and enterprising citizen.*[1]

The lungs of the Earth

There is no life for animals, plants, fish or micro-organisms without breath and breathing. Vegetation has a unique role in stabilising the atmosphere so that there is sufficient oxygen and not too much carbon dioxide. All vegetation on the surface of the Earth has this role but in the tropical rainforests the role is greatly magnified. Every single hectare of rainforest takes from the atmosphere one tonne of carbon dioxide per year. This alleviates the rising carbon dioxide concentration from the burning of fossil fuels, which in itself is causing global warming. Conversely, our present annual rate of deforestation of 150,000 square kilometres – two per cent of the Earth's tropical rainforests – puts two billion tonnes of carbon dioxide into the atmosphere each year. On this basis alone forests are essential to our sustained health and well-being. Nevertheless forests are being cleared at such a rate that they will be gone in a few decades and there will be an acceleration of global warming with all its consequences. (This is further discussed in Chapter 4.)

Forests have other beneficial effects on climate. They

breathe out or transpire a large amount of water as water vapour, thus cooling their own environment and that of the Earth; the forest creates its own moist climate and this causes rainfall. In turn rivers are created which supply water to millions of acres for food production throughout the world. Under some circumstances even native vegetation in parched regions of Australia has a similar role. Areas of native vegetation interspersed with cleared land may induce convectional currents which in turn lead to rainfall. The leaves of vegetation also induce precipitation by 'mist drip'. The mechanism is one of condensation onto the surface of the leaves from humid air or mist.

Forests are also essential as a vast filtering system for our water catchment areas and to stabilise our soils from erosion as well as protecting thousands of species (see page 72). Finally one has to admit, of course, that forests do have significant economic value under today's value systems. This is why they are plundered. But if the use of wood from the forest is selective and sustainable, the forest can be of permanent economic value. World Bank figures show that tropical timber exports throughout the world have declined from eight billion dollars in 1986 to six billion dollars in 1991 and are expected to fall to two billion dollars by the year 2000 or shortly thereafter. This is because forests are a short-lived and disappearing resource plundered for expensive dwellings and chopsticks and woodchip. The end result will be that we will be without part of our 'lungs'.

Deforestation

Deforestation is not new. Historically the temperate forests of the world have been destroyed. Scotland has become bare hills, the Mediterranean rocky and eroded. The trees have gone. The reasons were many but were not substantially different to the plunders of today. Until AD 800, the Scottish highlands were covered by a great forest of oak, birch and Scots pine with a rich wildlife that included the brown bear and the wolf. Large areas of forest were burned by the Vikings to 'flush out' the natives and the destruction was continued during feuds between the Scottish clans. During the sixteenth century forests were destroyed for smelting and in the twentieth century for packing cases used in the two wars with Germany. Little remains of these forests. The cleared land was turned over to sheep and the ecology of Scotland changed forever. In the Mediterranean region it was agriculture that caused the clearance. At first this was localised, then as the population grew, all the forests were cut to make way for crops. The soil deteriorated, sheep replaced cattle and then goats replaced sheep. Today only the goats are retained on the bare, rocky hills.

The deforestation taking place today is primarily of rainforest for most of the temperate forests have gone. The pressures on rainforest arise from a variety of sources. Paul Harrison indicates that population growth in developing countries accounted for 79 per cent of deforestation between 1970 and 1988 as poor people sought to eke out an existence from personal farming.[2] In other parts of the world, deforestation continues as part of war or colonisation; vast tracts of Tibet are being denuded by the Chinese leading to erosion and flooding in the countries of South-East Asia which lie downstream. War, as in Cambodia, makes this plunder easy and cash is always necessary to buy arms. Many governments of developing countries participate in the plunder because of debt imposed by international institutions and Western countries. Their attitude is that Western countries have chopped down the trees in their countries; why shouldn't they chop down their trees to provide health services and infrastructure? Or, in some cases the rulers are simply corrupt and gain financially from the plunder.

But the major culprits are the large international companies which often hide and disguise their activities under other names and which plunder for profit without any consideration of local inhabitants or the future of the world. Our economic system has made our governments beholden to these companies and little is said or done to stop them. Their image is fashioned as good corporate citizens in the advanced countries. The most unforgivable actions are surely in the most wealthy countries which do not need to burn and plunder. Let's look at two of the most wealthy of them, Japan and Australia. Japan plunders native forests throughout Asia, Australia plunders its own.

Japan and native forests

Japan is the world's major importer of tropical timber. The imports amount to 30 per cent of world trade in this timber. In Japan high-grade hardwoods are used by the construction industry and for plywood and much is discarded after short-term use. These practices have continued because they are economically viable to the Japanese – a pittance is paid to the peoples and governments of poor countries for clear-felling of native forests. The trade is run by subsidiaries of some of Japan's major companies which lie at the heart of Japanese economic success. The destruction of this resource in South-East Asia has had terrible consequences for there has been little attention paid to sustainability. The economic benefits to those countries have been minimal and in many places indigenous peoples, the forest dwellers, have been forcibly removed or have been displaced because their source of food and living has been destroyed. In more recent years Japan has been joined in its plunder by other countries that have undergone rapid economic growth, such as South Korea and Malaysia.

Australia and forests

What role should Australia be playing in preserving the lungs of the Earth? The lucky country now aims to retain 15 per cent of the natural forests that existed at the time of colonisation. The battle for the forests is vigorous and continuing. Government focuses on the short term and is more worried about jobs than the future. It needs to worry about both equally. Education about the importance of the environment doesn't seem to have made much difference. In the state of South Australia the far-sighted Native Vegetation Act was passed nearly two decades ago to preserve the state's remaining vegetation. Prior to that there had been tax incentives to clear vegetation. Governments, prominent politicians and companies try to evade the act. However, the situation is far worse in other states, particularly Queensland, where there is massive clearance of Brigalow scrub (the native vegetation that had once covered all of inland Queensland) from marginal lands. If an educated, wealthy community cannot appreciate how we breathe, then what hope is there for humanity worldwide?

The situation in Australia is an embarrassment to thinking people. Independent scientific reports indicate that the woodchipping industry, which commenced in the 1970s, is not sustainable and it breaks the Australian federal government's own principles of sustainable development. We have state governments colluding to destroy areas of forest that are subsequently proven to be of World Heritage value and listed as such. We have major national companies spending millions on advertising the benefits of chopping down trees because this offers such an immense profit. Subsidies are provided for putting roads into forests and for diesel fuel for the machinery, with the result that a profit can be made by woodchipping even though the export price is low. As for the jobs, they are steadily reduced in numbers due to mechanisation whilst the amount of woodchip exported increases. The defenders of the forests are persistently urged to compromise in the interests of jobs and economics. Historically, compromise has always meant 'let's chop down a little more'. The destruction is relentless.

Profit provides the motivation for logging companies to manipulate the minds of the population and to bend the scientific evidence. This quasi-science talks about 'normalising' old growth forests. 'Normalise' means 'cut down' so that the regrowth trees will all be the same size and easier to harvest next time. Company representatives move into areas of old-growth forests that are privately owned and tell their owners that the forest is now unhealthy since there are no saplings to be seen. They are then told that the forest will return to health if the old trees are cut, thus allowing light to come in and new trees to grow. Today, fortunately, most school students understand that an old-growth forest is an integrated biological system with species changing slowly over thousands of years. An important part of this cycle is that there are old trees that either die or fall down and during their disintegration they are an important part of the habitat for perhaps the majority of species in the forest. These species depend upon hollows and holes in trees, and the nutrients from rotting wood. If the tree is actually blown over or falls over, then and only then is sufficient light provided for seeds to germinate and successfully produce new trees. This dictates that the natural renewal rate of mature forests is extremely slow. Cutting selected trees from old growth forests disrupts some biological processes with loss of habitat and loss of nutrients. So harvesting is not a controlled extension of the natural process and if we are to retain biodiversity we must retain a spectrum of old growth forests. All these facts are disputed by the forest industries. A most interesting phenomenon is the development of societies such as Australia's Forest Protection Society, which is run for companies and loggers. It reminds us of the 'double-speak' of George Orwell's *Nineteen Eighty-Four*, in which a totalitarian regime used the opposite meaning of a word to bend the minds of the tame populace.

However, not all is totally lost in Australia. The Daintree forests, a remnant of the more extensive rainforests, were saved by environmental activism and were World Heritage listed. Before the listing, logging was valued at 30 million dollars per annum. Now, eco-tourism brings in double this amount each year. But other temperate forests in Australia of equal importance to the world are being woodchipped. Fortunately Australia is a democratic country in which activism can exist freely, whereas in many other countries forest activism results in prison or worse.

The major environmental health consequences of deforestation are therefore easy to see. Global warming will increase, rainfall will decrease over the denuded areas and rivers which provide water for drinking and crops will yield less. By contrast, some rivers will be responsible for more dangerous floods when rainfall does occur because water is normally trapped by vegetation and then released slowly. The catchment areas so necessary to our public health (because of our need for clean water in urban areas) will also decrease: their hillsides will remain denuded and soil erosion will increase. The eroded soils will be washed into wetlands, destroying their ecology and stopping their use as a source of water for crops. All these events have occurred in Cambodia where warring armies have sold the timber to buy arms. There is now frequent flooding, massive erosion, loss of wetlands, and falling food production.

Deforestation will be responsible for more subtle effects on our health. Since the forests, particularly the rainforests, carry a huge number of species, disturbance to their web of life will have the greatest consequences. We will explore the origins of two ecological diseases, one of which emerged from the forests of New England, USA, and the other from the forests of Central Africa.

Like many of their fellow Americans, Jim and Dale loved camping with their three children. It offered the opportunity for adventure, close family contact and friendships with other campers.

They were happy not to travel far so would go into the hills, pastures and woods near their home in Hartford, Connecticut. Three days after one holiday, Dale developed a headache and fever and was soon languishing in bed with 'influenza'. She also had some red blotches on her skin, but these were thought to be due to the high fever and in any case they soon resolved. Her fever was still present after ten days and she felt very unwell. Then one morning she awoke with her heart pounding in her chest and a feeling that she could not breathe. She was taken to hospital as an emergency. There it was recognised that her symptoms were those of an infection called Lyme disease and the diagnosis was confirmed by a blood test. However, by this time the infection had attacked her heart, leading to palpitations and heart failure. She required intensive treatment and antibiotics given into her veins before she could recover.

Whilst camping Dale had been bitten by a tick which had transmitted an infection called Borrelia to her. The first epidemic of this infection caused arthritis in the village of Old Lyme in Connecticut, but in some cases it can cause inflammation of the heart or the brain. Lyme disease is caused by man changing the ecological balance of species in the forests of New England. The Borrelia infection is carried by mice. The ticks on deer in the forests feed on these mice and in so doing become infected themselves. When the ticks bite humans the infection is transmitted. The populations of mice and other rodents have increased greatly in recent decades because farmers cleared forests and killed wolves and coyotes which had killed the mice and deer. More deer and mice meant more ticks. As people went into rural areas more, exposure to ticks was more likely. Some 10,000 cases of Lyme disease were reported in the USA in 1991. Similar infections are also widespread on the European continent.

We can learn from this disease that altering the structure of ecological communities can have unforeseen consequences for human health, either directly or indirectly and sometimes not for many years. Diseases transmitted to humans by ticks occur in many parts of the world. The presence of ticks is likely to be affected by global warming (see Chapter 4) and some of the diseases caused by them will increase.

Tampering or interacting directly with 'untouched' ecosystems is potentially dangerous. There is evidence now that infections of great virulence reside in complex tropical habitats. Incursion into these habitats on a large scale, coupled with rapid movement of people, could contribute to the emergence of such lethal epidemics as AIDS. It is only by good luck that the HIV virus, which is the cause of AIDS, is not transmitted more readily than it is. If, for example, the virus which causes AIDS evolved to be transmitted in the droplets of moisture breathed out or coughed by humans, then there would be a massive epidemic. The pneumonic form of plague transmitted itself in this way. The infection of plague attacked the lungs and was spread to other persons by coughing.

Is AIDS an ecological disease?

Identified only in 1981, the HIV virus had infected ten to 15 million people by 1992 and will have infected 50 million by the year 2000. HIV has many of the hallmarks of an ecological disease. Some scientists suspect that the virus has transferred to humans from another species, perhaps another primate (monkey) in Africa, under conditions of rapid environmental change and social disruption, basically brought about by the imposition of Western economic systems in greatly 'underdeveloped' counties – that is underdeveloped in Western economic terms, but not necessarily so in social or cultural terms. In Africa, the traditional tribal agriculture of mixed farming and gathering were rapidly replaced by the cash crop system to pay for loans given for inappropriate 'development' projects;

native vegetation was cleared, displacing species and possibly bringing man into contact with new infectious agents. The replacement of tribal culture and rules by a Western-style economic free-for-all led to new employment patterns (or perhaps non-employment patterns, with widespread search for work) and increased promiscuity, which fostered transmission of the virus. The virus then spread through the ecological deserts of our 'developed society' – New York, Los Angeles, Bangkok, and various European cities – encouraged by social conditions and increasing intravenous drug use, one of our most profitable economic developments. The HIV virus may be the first virus to harness rapid ecological change in its widest sense – rural, social and urban. It may even represent a natural counterbalance to our population growth and destructive activities.

If this theory for the origin of AIDS is correct, we should learn from the prime causes of this disaster and not repeat our ecological mistakes. There is another theory on the origin of AIDS. Cells from the kidneys of monkeys were used for the production of a polio vaccine subsequently used in Africa. It is possible that the vaccine contained the HIV virus. If this is so, the environmental and economic factors described above led to its spread.

Outbreaks of *Ebola* virus disease are also occurring in communities near to tropical rainforests in Africa. The virus infects humans, probably from monkeys. It causes a severe fever and bleeding into the skin and internal organs. The disease has an horrific reputation because the body appears to disintegrate or dissolve rapidly and most patients die. There is speculation that this virus, like the HIV virus, has emerged from the rainforests as a result of rapid ecological changes. Of course, infections have managed to jump from an animal species to people many times during human evolution, but rapid ecological change, together with overpopulation of humans, is likely to increase this potential.

As well as viral infections such as HIV and Ebola, other infections, particularly parasites, may emerge from the destroyed forests. In

Central and South America the bugs that carry the infection American trypanosomiasis, an infection by the protozoa *T. cruzi*, have moved into thatched houses and more infection now occurs in humans. The bugs emerge at night to feed on the occupants, on whom they defecate. The faeces, which carry the infection, are rubbed into the eyes, mouth or cuts on the skin then spread by the victim's bloodstream. The infection destroys the heart, the gullet and intestines.

Attitudes, politics and deforestation

A headline in *The Australian* newspaper not long ago proclaimed 'Forests threaten development'.[3] This illustrates today's economic attitude to the preservation of forests. The vision is of forests marching forward to impale or crash onto human endeavours such as houses and factories. The article explained that 'investment projects worth about two billion dollars in Western Australia were threatened by the latest proposals designed to save forests – according to senior WA government officials'. In other words if the remnants of forests in Western Australia were retained, and indeed they are only remnants now, proposals put forward by industrialists including a pulp mill which would no doubt wipe out most of the remainder would have to be forfeited. And this is not an isolated case. Look around you: you will find a neighbour who will cut down a swathe of 100-year-old trees to give himself a better view from his house. Along the coastline, if a prime view of the sea is interrupted by trees, they will not last for long. Anything that affects our short-term economics or needs will go and anything of quick value to us individually, collectively or company-wise will be plundered.

This is also an issue in each city and town on this planet, for we need to breathe and we need clean air. We must think not only about forests but also about the greening of cities. The trees in gardens and along freeways act as sinks not just for the carbon dioxide but for many other pollutants issued in our industrial society. Sinks are just what you imagine them. They absorb or drain a pollutant from the environment and convert it to a non-pollutant, they convert carbon into wood, and change carbon dioxide into oxygen. We also need to think about our farmlands. Though greatly modified by man, these lands still contain trees. Yet in the farmlands of the great open areas in Australia and other countries, a tree in a paddock is an enemy. It stops the rolling efficiency of farm machinery. It doesn't seem to matter too much for the moment, for the next year, or the next five years that millions of tons of topsoil will fly over the cities and into the oceans, blown by the wind from treeless paddocks, as long as the profit is made.

The action of destroying forests in healthy countries such as Australia has an important implication for forests worldwide. The destruction of forests in South-East Asia is massive and not sustainable. In the Solomon Islands, Papua New Guinea and the larger countries, logging concessions are obtained for a pittance, indigenous peoples are displaced and cleared land left to the mercy of erosion. The root cause is the same in both rich and poor countries – the greed of the large company and the individual. Criticism or advice from wealthy countries seems hypocritical, given that the same devastation is occurring in developed countries. A Brazilian scientist, responding to advice from a scientist from a developed country, said:

> I expect you're just like the scientists and the bureaucrats from the World Bank who come here. You are all in Brazil trying to stop us exploiting the wealth of the Amazon. What arrogance! You've already destroyed all the forests in your own nations and so you come and lecture us about the environment ... We're not going to listen and we're going to develop the Amazon.[4]

Biodiversity

We come now to a difficult concept, biodiversity. The term biodiversity is used to describe the great variety (diversity) of species and therefore refers to all living things on Earth from microbes to the largest mammal, from algae to the largest tree. In Australia, the importance of maintaining biodiversity has been recognised by the establishment of a National Biodiversity Council of eminent scientists, which has criticised the government's forest policy. It is maintained that the policy will not protect biodiversity. Preservation of biodiversity is linked with the preservation of tropical rainforests because they contain a vast number of species. These forests act as the powerhouse for the evolution of species and contain half of all the species on Earth. A small plot of rainforest will have 700 species of trees compared to the handful of species to be found in other woodlands of the world, and thousands upon thousands of insect species, many of which are unidentified. This huge genetic bank of knowledge is being wiped out as the forests are cleared – most of the species will become extinct for they do not exist elsewhere. We do not know why so many more species have evolved in these rainforests than elsewhere. Presumably all necessary conditions of moisture and temperature have been present for millions of years. And we should remember that destruction of the forests destroys not only the species living there but it also prevents the birth of new species that would benefit from the conditions.

We will now explain why the retention of species is necessary for the future health and well-being of our species, humans. Three arguments have been put forward for the retention of biodiversity. Firstly, it has an economic value: it provides a breeding stock for plants and animals and it is a source of pharmaceuticals and chemicals. Secondly, retention of biodiversity is an ethical issue related to our place in nature and the recognition that humans are part of the ecological web of life. Thirdly, as we have seen in our discussion of the importance of forests, biodiversity is important in maintaining our physical environment, temperature, rainfall and the purification of wastes and pollutants. This latter argument is discussed elsewhere in this book. Here, we discuss the first two arguments and explain their importance to our health and well-being.

The economic value of biodiversity

The Earth's forests, seas and soils can be thought of as a factory that produces millions of chemicals and pharmaceuticals for our use. It would be impossible economically for the pharmaceutical industry to perform such an extensive creation of substances. Between a quarter and a half of all drugs used in Western scientific medicine come from nature and in developing countries nearly all health care is provided by traditional and natural medicines. There is an increasing recognition by the pharmaceutical industry that many herbal remedies contain cures which can be researched and copied in the laboratory. Indeed, longstanding herbal remedies would be expected to contain important pharmaceutical secrets for their use has been honed successfully over thousands of years. For example, a traditional Chinese remedy contains substances that are effective in the treatment of Hepatitis C. International pharmaceutical companies have recognised nature as a potential source of lucrative finds by paying countries for permission to search their forests. These facts have wide implications for the continued success of Western scientific medicine. But more importantly, we should realise that these discoveries represent the successful fruits of ecology, the balance of nature. The leaf of a plant produces a chemical to deter or kill an insect which eats it; a fungus produces a chemical (for example, an antibiotic) which kills the bacteria near to it and so allows it to survive. So somewhere in nature lie the chemical secrets and structures which will help us deal with many more human diseases, including today's epidemics such as AIDS.

Many of the most important and widely used drugs in Western scientific medicine come from nature. Aspirin, used worldwide for pain relief and to prevent blood clotting in patients with heart disease, is derived from the bark of the willow tree. The bark from the cinchona tree in Peru produces quinine for the treatment of malaria and most newer drugs for malaria are derived from its chemical formula. The bark of the Pacific yew has produced Taxol, probably the first really effective drug therapy for ovarian cancer.[5] Presumably trees produce all these substances in their bark as a defence against damage. 'Defensive' chemicals are also produced by the leaves of plants and trees, and chemicals called alkaloids, found in the periwinkle, have given us effective treatments for childhood leukaemias. Fungi produce a variety of chemicals to protect themselves from the millions of bacteria in the soil; a fungus produced the penicillin discovered by Howard Florey and other fungi have led to the discovery of the new antibiotics, the cyclosporins. Nature has produced drugs that help us with anaesthesia, childbirth, heart disease, high blood pressure and Parkinson's disease. The list is endless and will stretch into the distant future if we are prudent.

A second, equally important economic argument for retaining biodiversity is our food supply. Worldwide, there are probably more than 250,000 species of plants, yet as explained by Richard Leakey and Roger Lewin, our selection and breeding of fewer and fewer species has gradually resulted in only 20 species providing 90 per cent of our vegetable food.[6] Three species of cereal provide half the world's food. Vast areas of the world are covered by just one crop.

A monoculture is one species selected for its productivity and grown to the exclusion of all other species. A dependence on monoculture is potentially disastrous to humans, largely because the entire monoculture can become susceptible to disease, thereby endangering food supplies. The potato famine in Ireland in 1845 is a good example. A fungal disease, encouraged by wet, cold weather, destroyed the potato crop for two successive years. There were few other crops and the population starved. Of a population of eight million, one million people died and two million were forced to emigrate. An ecosystem that contains only a few species is always vulnerable to the invasion by new species as was the human colon after the use of antibiotics. By contrast, a more complex agricultural system such as mixed farming reduces the risk of pests and weeds. Just as the spider's web is stronger the more cross strands it has, the more species there are in an ecological community, the stronger that community is. Scientific studies show that grasslands with many species are more resistant to drought than those with few species.

An interesting example of the ecological effects of monoculture comes from Queensland, Australia. Sugar cane is an important crop and to facilitate harvesting, the fields were expanded by removing trees and hedgerows. Ten per cent of the crop was then being eaten by native rats, which had proliferated in the grassy corridors between the cane fields. Poisoning the rats proved futile for the poison also killed the natural predators. So trees were replanted, which deterred the growth of the grass and reduced the cover for the rats. At the same time owls, which previously had nested in the holes in the trees between the fields, were encouraged to return. Tall metal poles with a horizontal perch at the top were erected. The owls perched on them to sight and attack the rats. Clearly it would have been simpler to retain the original ecological system.

We need to be able to renew the vigour of our plants and stock by introducing genetic material from 'wild' species. We must acknowledge that the world will reach a population of about ten billion by 2050; the demands on natural resources for commercial and subsistence purposes will reach levels never known before, yet the fact is that we have identified only about 15 per cent of the species we live with on the planet and we know even less about what they do and how they interact. We should be very resistant to continuing forms of development which harm the world's biodiversity.

The authors, however, disagree with the narrow economic argument for the retention of biodiversity, for there are great dangers in having economic growth as the world's unofficial religion, as is the trend at present. The argument that species have value plays into the hands of the economic rationalists in a world where nothing has value unless it can be traded or utilised. Putting a price on biodiversity can lead us along many dangerous paths. In terms of the discovery of drugs, the computer analysis of potential novel structures discovered from nature will expand, so that economists will come to believe that we have recorded millions of options. They will say, 'Why retain this or that part of nature when instead we can "develop" the land?' 'This forest's biological diversity is worth one billion dollars, why retain it when we need a satellite launching station on the site which will be worth ten billion?' The authors conclude that it is too dangerous to sup with the devil – human greed.

We are part of it

There are aspects of human life that cannot always be rationalised. The great mountain climbers of the world have never managed a better explanation of why they climb Everest, K2 or the Matterhorn in winter than 'because it's there'. It is not just that it is a challenge – humans are challenged by many situations; it is because it is an emotional experience or a wish to experience the natural word.

The ecologically aware human believes that biodiversity must be maintained because we are part of this web of life and perhaps, more difficult to understand, because biodiversity is part of us. Those who cannot accept this argument can accept biodiversity as a support for our future through new genetic material for future drugs or cereals. But once the discussion extends beyond this, the conservationist is accused of being emotional or irrational. By the same token of course, all religions can be seen as emotional and/or irrational yet we do not dismiss them so lightly. It should be recognised that for many humans a healthy life has the dimension for an ecological life, one that allows us to fit into the web of life so that we have meaningful relationships with other humans, nature and the land.

In *The Fallacy of Wildlife Conservation*, John Livingston argues that we have to analyse the belief systems that are developed in us during our childhood and education.[7] He maintains that we are not born 'rational' in our outlook; rather, it is developed and taught. There is a division in our lives between being rational by using and applying scientific methods, and being non-rational by accepting mysticism or religious experience. We separate intellect from emotion, we separate reason from feeling. Our society demands that economic and organisational matters are dealt with by rationality and reason. We develop bridges between rationality and non-rationality and some of these bridges are given approval by the community by being given such names as 'culture'. The title of his book indicates that Livingston believes that wildlife and its biodiversity cannot be dealt with by argument or rationalisation, it is something akin to religion in that you either accept it or reject it. Livingston points out that what he calls wildlife preservation, – the maintenance of biodiversity – is entirely dependent upon individual human experience; this cannot be measured or argued over any more than the human experience of colour or religion can. Such experiences of nature have been described by many and by those who have seen the Earth in a different light, such as astronauts who view the beauty of the Earth as a single organism from space and return to Earth as changed individuals. Livingston describes vividly how as a ten-year-old he became immersed in the life of a community of toads, frogs and newts near his city home:

> *The longer you looked, the more deeply you were mesmerised . . . possessed. There was no world whatever, outside that world . . . nothing beyond shimmering light on water, smooth clean muck, green plants, trickling sounds, flickering tadpoles, living, being. Plans were revealed for the construction of a storm sewer through 'my' ravine. Shock, dismay, and all the rest of it were mine early. The ten-year-old mind is not subtle: how can I warn the frogs and toads and newts? Can I get them out of there, take them away somewhere? They are defenceless; it is wrong to hurt them. What right do we have to hurt them when we cannot warn them? They don't know what is happening, or why. There was much puzzlement here. All logic seemed to be backwards or upside-down; nothing made sense. I could do nothing but watch, with sorrow and fury.*[8]

As Livingston says, he was responding this way because he had become part of the creek and he was being damaged. His conclusion was that there can be no rational or scientific argument for preservation of biodiversity (wildlife). Yet the philosophy of our civilisation is increasingly economic, competitive and geared to self-aggrandisement in economic terms. The ideas of fundamental 'Greens' represent a completely new philosophy of life, a philosophy which will conflict with those individuals who wish society to continue as it is at present. Green ideas are therefore subversive and will arouse frustration and indeed anger amongst those who are prevented from gaining their ends. It is a conflict of ideas that will increase as the world battles for sources of water, land and the fishes of the sea.

The economic argument is always based upon expediency – growth is needed now, so we must cut down more trees or we must develop this mine despite it being in a national park. It is not based upon future long-term considerations and the possibility of adverse effects on health and well-being. Even in the field of environmental health we blunder ahead with so-called essential developments before their implications are understood. The mobile phone mania is a case in point. The development is explosive, it happened in a few years, a supposed aid to economic growth. Only now when millions of phones are in operation is there any public concern about brain tumours in users. In the sphere of ecological health, it is clear that warnings of possible future disasters are even less likely to lead to government action when the future problems are 30 years hence and cannot be predicted precisely. This issue is well illustrated by the phenomenon of global warming.

---

1. Henry David Thoreau, quoted in *Wilderness News*, Glebe, New South Wales, Aug/Sep/Oct 1995.

2. Paul Harrison, *The Third Revolution: Environment, Population and a Sustainable World* (I.B. Tauris, London, 1992).

3. *The Australian*, 16 August 1995, Business Section.

4. 'Into the Forest', *New Scientist* Vol. 151 (21 September 1996), p. 26.

5. Taxol is a registered brand-name of Bristol-Myers Squibb Pharmaceuticals.

6. R. Leakey and R. Lewin, *The Sixth Extinction* (Weidenfeld & Nicholson, London, 1996).

7. John Livingston, *The Fallacy of Wildlife Conservation* (McClelland Stewart, Toronto, 1981).

8. Ibid.

CHAPTER 4

TURNING UP THE HEAT: GLOBAL WARMING, ENERGY USE AND HEALTH

Who would expect thousands of inhabitants of London, Melbourne, Paris and Adelaide to be infected by malaria during the next century? Global warming is occurring and with each year that passes without preventative action we are condemning future generations to higher temperature rises, more prolonged warming, greater climatic instability and to diseases such as malaria.

With global warming, many of the unconquered diseases of the underdeveloped tropics will become common in Western communities. We know this because their spread is already occurring. Computer modelling, accomplished by feeding into the computer the present-day worldwide temperatures, rainfall and humidity, confirms what we are observing. If the global temperature rises by one degree Celsius then the change in climate at any one place or city in the world can be predicted. At the same time we know the temperature and humidity required for a particular infection to thrive and the computer will tell us about their spread. This chapter will look at some of the expected changes in health and disease due to global warming.

Jason, a New Zealander, was an avid adventurer, enjoying the challenge of travelling alone to the forests and the huge primeval rivers of South-East Asia. For two years he worked as an engineer on a mining project in the mountains of Irian Jaya. After that he had two months leave and decided to trek north over the mountains and descend into the river valleys which flowed north to the sea. The forest was impenetrable except on hidden tracks used by the indigenous peoples. He employed villagers to show him the way. The heat and humidity became terrible as he gradually descended, and it was usual to be exhausted and hungry.

He had done his homework on health hazards before this adventure. To prevent infection from malaria, he took a tablet of Larium each week, he used insect repellent freely and slept with an effective mosquito net around him. After four weeks the river he was following widened and he purchased a dugout canoe from the natives to speed his progress. At times the air was black with mosquitos and they seemed capable of biting through anything. After two weeks on the river Jason felt he had 'flu but then suspected the water he was drinking because he developed headaches, nausea and some vomiting. After a few days he realised he had fever. Despite the daytime temperature of ninety degrees Fahrenheit and 100 per cent humidity, he felt as cold as ice and shivered and shook uncontrollably. He went ashore, pitched his tent and rested. By now he realised he had severe fever but could scarcely accept that he might have malaria – it seemed inconceivable when he had taken his prophylactic Larium so diligently.

After a week of high fevers, which seemed to come on every second night, he lay in his tent in a small village, ill, debilitated and exhausted, with the natives bringing him fruit and water daily. They knew the nature of the problem for they had seen him look ill and heard him calling out in his sleep. Fortunately they contacted a government station 20 kilometres away where a phone was available. Two days later he was evacuated by air and was treated for two forms of malaria, vivax and falciparum. He recovered.

Jason had adopted all the correct measures to avoid malaria on his solo journey but the amount of infection pumped into him from the bites of thousands of mosquitoes along the rivers overwhelmed the defences of the drug he was taking. In the areas of Irian Jaya he visited, malaria is endemic – everyone has it, but most natives have built up sufficient immunity to remain alive, even though their health is impaired. Most of the tropical regions of the world have endemic malaria. According to the World Health Organisation, 2700 million persons – approximately half the world's population – are exposed to malaria and 270 million are infected. Each year two million people die from the disease, mainly children. More so than with any other disease, the risk of contracting malaria is decided by the local environmental conditions: the temperature, rainfall and humidity which determine the presence of mosquitoes in sufficient numbers, also determine whether malaria will be present. We will return to the topic of malaria to explain how and why it will spread once we have discussed the cause of this spread – global warming. The spread of malaria will result from ecological changes and it is therefore another example of an ecological disease.

In 1996 a report was prepared on *Climate Change and Human Health* on behalf of the World Health Organisation, the World Meteorological Organisation and the United Nations Environment Programme. The report begins:

> *We are living at a remarkable moment in the history of the human species. Human population size, and the extent and nature of our economic activities are now so great that the gaseous composition of the lower and middle atmospheres has begun to change. This is likely to affect the world's climate, many other of the world's natural systems, ground level exposure to ultraviolet radiation, and indeed all life on Earth.*[1]

The scientists are saying, whilst avoiding causing undue alarm, that climate change due to global warming is an immense threat to the survival of society as we know it even over the next 100 years. Our psychology is to find it interesting but not worrying.

The problem for mankind is that once we experience the effects of global warming, nothing we can do will immediately change the situation. The effects will be extensive and devastating and will go on for centuries, even if corrective action is initiated. Global warming links with everything else we discuss in this book, from the destruction of forests to our economic system, multinational companies and political processes.

The Greenhouse

Why do some of us own a greenhouse? Because the air inside stays warmer than the air outside. Why is this so? Radiation from the sun comes to Earth as short-wave energy. When this passes through the glass or plastic cover of a greenhouse, it is absorbed by the air, plants and earth in the greenhouse and forms heat. This heat is re-radiated upwards as long-wave infra-red radiation and is trapped by the glass or plastic and retained within the greenhouse. The same process creates a natural greenhouse effect for the Earth. Instead of the glass or the plastic of the greenhouse, carbon dioxide, which is one per cent of the Earth's atmosphere, traps the heat radiating from the Earth and prevents it being lost into outer space. If we wanted to heat up our household greenhouse we could thicken the glass or the plastic.

In the same way thickening the layer of carbon dioxide in the atmosphere will increase the heating of the Earth. This carbon dioxide content of the Earth's atmosphere has kept the temperature of the Earth relatively stable over millions of years. Fluctuations have been very slow, allowing plants and animals to adapt to any changes. Scientific data from the study of ice cores in the Antarctic indicate that, over long periods of time, a slight increase in the carbon dioxide in the atmosphere increases the Earth's temperature slightly. The cores of ice are taken from a glacier formed by the accumulation of layers of snow over 160,000 years. Throughout this time, despite several ice ages interspersed with warm periods, carbon dioxide in the atmosphere ranged between 190 and 280 parts per million. But the ice cores show that, since the industrial revolution 200 years ago, carbon dioxide in the atmosphere has increased by almost one-third, from 280 to 360 parts per milion. Half of that increase has occurred since the 1950s.

In scientific studies it is always important to have confirmatory studies. When a tree is felled, the number of concentric rings on the cross section of the trunk indicate the age of the tree. But the width of each ring tells us about the climate of the year in which the ring was produced. A narrow ring means a colder year and a wider ring, a warmer year. Tree rings can be measured over 10,000 years by studying trees buried in volcanic eruptions and other catastrophes. Such studies confirm recent significant global warming.

The increase in carbon dioxide that leads to global warming is caused by the burning of fossil fuels. However, other gases contribute to a lesser extent, for example, methane. This has increased even more rapidly than carbon dioxide. Methane arises from rice-fields, landfills and from the guts of cattle. It may also arise from the soils of the Arctic regions which are warming sufficiently to release their store of methane. Global warming will establish a vicious circle whereby increased methane is released.

Science and the greenhouse effect

For over a decade conservationists have been expressing their worries about the potential greenhouse effect. Initially their claims were derided and much of this derision came from those powerful voices with money to lose if industry and commerce could not proceed as usual. Multinational corporations (particularly those producing electricity), mining companies and economists continued to run the world along the lines of 'business as usual'. 'Independent' institutes were set up, funded by business, where economists and scientists could refute the claims about greenhouse gases. And even if they could not refute them they passed judgements, claiming that the gains might be greater than the losses if the temperature of the world rose. Even if we lost a Bangladesh owing to the increased sea level there would be areas of the world that would become more productive in terms of food production, so why worry? There is a common belief that humans can always overcome, alleviate or deal scientifically with whatever problems nature creates.

Mainstream science now accepts that global warming is occurring. In 1988 an international panel on climate change (IPCC) was established by the UN Environment Program. This expert body of over 300 scientists has now provided estimates that the global temperature has increased by 0.3–0.6 degrees Celsius over the past hundred years. The most recent estimates indicate that by the year 2100 the surface temperature of the Earth will have increased by between one and 3.5 degrees Celsius. A three degree rise would make the world much hotter than it has been since the emergence of humans two million years ago. The consequences of this are horrendous. However, even these estimates may be underestimates as we will see from the discussion of the balancing mechanisms that will determine global warming.

The carbon cycle

Before we burnt fossil fuels the main sources of carbon dioxide in the air were natural fires, the breathing out of plants and the decay of matter. Carbon dioxide concentrations did not rise in the atmosphere because carbon dioxide was removed by two main methods. Firstly, photosynthesis in plants whereby, with solar energy, carbon dioxide is built into the carbon of the plant. Most effective in removing carbon dioxide are the forests of the world. Secondly, carbon dioxide is absorbed into the oceans of the world; these also act as 'sinks'. Many scientists believe that their capacity to remove carbon dioxide is massive and this will prevent global warming. Some recent evidence, however, suggests that this claim is spurious and that the oceans may not be able to cope at all with further significant rises in carbon dioxide. If this is so a rise in the Earth's temperature could be catastrophic because it will be much greater than the increase of 3.5° predicted by 2100.

The physical effects of global warming

It is predicted that by the year 2100 the sea level could rise almost 100 centimetres, having already risen by 10–25 centimetres in the past 100 years. Thus vast areas of the world that are just above sea level at present would be flooded and then submerged, including many small nations of the Pacific, large parts of Asia and significant parts of other continents. River deltas form 80 per cent of the land of Bangladesh. Today its 110 million people suffer regular floods from tropical cyclones; a rise in sea level of 100 centimetres would displace 11 million people. There are similar highly populated coastal deltas in China and Egypt. Displaced populations would have to be relocated and to date the world has been unable to find politically acceptable methods of relocating much smaller numbers of people. Some of the rise in sea level would result from the melting of ice in the polar regions and there would be massive disruption of the ecosystems in the southern and northern oceans which are the feeding grounds for much of the fish we use as food. This, together with the inundation of coastal mangrove swamps and wetlands, would be likely to produce enormous changes in marine ecology. In fact, a sea-level rise of 100 centimetres would threaten over half of the remaining coastal wetlands of the world. Another cause of the rise in sea level would be the expansion of the water in oceans due to the higher temperature. The temperature on land would also increase throughout the world, particularly in highlands; this would profoundly affect many parts of the world. Let us examine two simple examples.

In the cities of the world a rise in temperature would have serious effects on the health of the poor and the survival of those who are already ill. When the air temperature is greater than that of the human body (98 degrees Fahrenheit or 36 degrees Celsius), there is considerable stress on the body because to keep the body's temperature from rising the heart has to work harder and perspiration and sweating has to increase. In itself, this extra work for the body leads to an increased numbers of deaths, particularly among the young, the elderly and those with heart or lung disorders. Current predictions are that Washington DC, which now has on average one day a year over 38 degrees Celsius, will have 12 per year by 2050. By 2050 there will be more large cities with more crowded populations than today and many cities will be affected much more seriously than Washington. Studies in the United States indicate that civil disorder is common at these times. No doubt humans will find a way of alleviating the distress in hot cities, certainly those in wealthy countries will be able to turn to air conditioning – which will further increase the greenhouse effect. The poor people in developed countries and the universal poor of developing countries will have to suffer the consequences.

The second example is the effect of warming on the major river catchment areas of the world which in all continents extend from snowy mountain peaks to the river deltas of the plains. Rivers are fed with water from the mountains over a large part of the year.

This is because the mountains have a covering of snow in winter, which melts slowly over many months, feeding the spongelike systems of moss which gradually release the water into the streams, allowing them to run permanently and collect to form rivers. With a small temperature rise there will be no snow on some mountains and less snow on others; the rainfall of winter will have flowed down the streams and rivers in a few weeks, leaving the system relatively dry for the rest of the year. As a measure of these changes all mountain glaciers have retreated steadily this century. It does not require much imagination to work out the effects on ecology, agriculture, water supplies, hydro-electric systems and coastal wetlands. Apply these changes to the vast populations of Asia, which depend on rivers that flow from the mountainous regions in the north. The lowland regions of South-East Asia have already recorded a major increase in inclement weather including floods, typhoons and storms – an increase now acknowledged by the world's insurers.

The rise in temperatures will bring about many other harmful physical changes. While some regions will have increased rainfall which will help crops to grow, elsewhere much marginal land will be lost because of desertification. Climatic patterns will change in all countries so that rapid changes in agriculture will have to be instituted if drastic falls in food production are not to occur. Human health, in the broadest sense of the word, will suffer immensely from these changes which will lead to social disruption, hunger, famine, drought and flood. The number of natural disasters will increase and they will be more destructive. In Australia, there is emerging evidence that an increase in El Niño events are related to global warming. Since weather recordings began in 1877, El Niño (warm ocean surges of the Peru Current that cause the trade winds to cease) has occurred every three to five years. Changes in ocean currents and atmospheric pressure lead to drought in eastern and southern Australia and South-East Asia. There are grave implications for agriculture.

A precautionary principle, drugs and global warming

If an invention or development *may* have deleterious effects in the future, we should adopt appropriate precautions immediately rather than wait for them to happen. It is interesting to compare our attitudes to global warming with the development of a pharmaceutical drug. Both have the potential to harm countless individuals. Of course the threat from global warming is greater in that it could so damage the biological systems of the Earth that our way of life could never recover.

It is not possible for a pharmaceutical company to synthesise and market a new drug without extensive investigations into its safety. Indeed if investigations are not carried out thoroughly and according to the letter of the law, in many countries an undetected side-effect would lead to law suits and the likely demise of the company. These precautions have been adopted because we have learnt from experience that drugs may cause devastating side-effects, often in unexpected ways and to large numbers of the population. Thalidomide, which caused human birth defects, is one such example. Therefore, a new product is first screened, using mainly laboratory animals, for a whole variety of potential side-effects. Thousands of rats, mice and other animals die in this thorough process. In particular, the drug is tested for risk of damage to the foetus. Its effects on all the organs in laboratory animals are assessed by careful studies of the structure and chemistry of their kidneys, heart, brain and lungs. If all these exhaustive studies are satisfactory, the drug is given to human volunteers over a short period of time and its metabolism in the body intensively studied. Eventually the drug may be marketed for use in humans over short periods of time and if no side-effects are detected then it may be used for longer periods. Finally, even when the drug has been marketed, population surveys are conducted to determine if any unexpected side-effects are emerging. These exhaustive processes are to protect the individual and, as a result, the company. In fact the precautionary processes have become so exhaustive and of such long duration that there is community concern that the introduction of important drugs is being delayed (for example drugs for HIV infection).

It is interesting to compare this ultra-cautious process with the community's reaction – in particular, government reaction – to the commencement of global warming. If we used the argument that we have applied to the pharmaceutical industry, we would adopt immediate legal constraints on the discharge of carbon dioxide and particles from power stations and from motor vehicles worldwide. The cost of this would not be an issue in the same way that the cost of drug assessment to the pharmaceutical industry is not an issue. However, those sceptical about global warming argue that global warming has not been proven as yet. The authors believe that even if global warming was 'just a theory' the precautionary principle should still apply. Vast amounts of money are expended by the pharmaceutical industry even though it is unlikely that any one drug would cause cancer.

The real difference between these two situations lies in our belief systems (see Chapter 7) which place priority on the importance of the individual. If you buy a product which causes a side-effect that has not been predicted then you are harmed, your rights have been violated and you have a case against the manufacturer. Will a community of 3000 people on a Pacific Island that will be inundated completely in the year 2030 with the rise in sea level, have a case for recompense when this happens? Who would the case be against – the governments of 90 countries that have taken no action? The producers of power who have lobbied governments to take no action? The General Agreement on Tariffs and Trade (GATT) which has snuffed out alternative energy sources by refusing to allow their development to be subsidised? There are hundreds of possibilities but nothing can or will happen to assist the 3000 individuals whose island is being submerged. As we will see also in Chapter 7, it is not just a matter of individual rights being more important than community rights, it is a question of the human psyche and our inability to think about what will happen in 50 years' time. Unfortunately each of us is concerned only about our own lives and those of our family and this concern extends only for the next ten, 20 or at the most 30 years.

It is not always possible to use scientific data as the basis of all precautionary decisions, particularly when an event today has an effect in 30 years' time. This lack of understanding is common — take, for example, the following statement published in *The Australian* newspaper:

> *It is clear that concentrations of greenhouse gases are on the rise and that a causal relationship exists between such increases and increased global temperatures. However, the resulting impact on the natural or human environment is, at this stage, still a matter for research; indeed, as are the ecological and economic costs and benefits of such changes.* [2]

The problem with such attitudes is that by the time the experimental information is available, it will be too late to do anything about it.

The provision of energy

The community and industry demand electricity. This need and the way we handle it epitomises the entire environmental problem. The crux of the problem is as follows. Our energy is produced by coal, oil and gas which is converted into electricity. These fuels are sold cheaply and our energy is cheap – or is it? In fact oil, coal and gas are vastly expensive! Simply to maintain greenhouse emissions at the 1990 levels will cost industrialised countries 240 billion dollars annually for the next 20–30 years. If we added this cost to the present cost of electricity – just as the cost of developing a new drug is passed onto the consumer – the cost of our electricity would increase many, many times. We are deferring the real cost and are passing a debt to our children and their children.

The United Nations Framework Convention on Climate Change required countries to return greenhouse gas emissions to 1990 levels by the year 2000. Australia and many other developed countries will fail to reach this target. Yet the ultimate target will be to reduce carbon dioxide emissions to 30 per cent of 1990 levels if we are to stem global warming.

The position of Australia is particularly embarrassing. Australia maintains that its dependence on fossil fuels for energy and for export make it a special case; it should not have to reduce greenhouse emissions as much as other countries. It has made its international representations on the basis of information supplied to it by industry and commerce to the exclusion of independent scientific and environmental experts. By contrast, some international oil companies have responded to 'greenhouse' by deploying finance to the development of renewable energy.

There are many forms of renewable energy that cause no environmental harm, which are easy to develop. They have not been developed because the total cost of their production is slightly more than the current artificially low cost of electricity (nine cents per kilowatt per hour compared to six to eight cents). Of course, under a proper accounting system, renewable energy is considerably cheaper because it has no adverse environmental effects. Such forms of renewable energy are wind, solar and tidal power.

There appears to be a huge conspiracy against their development. The conspiracy starts with the multinational companies, which straddle the world, influence governments and act with impunity. These companies produce coal, oil and electricity and their profits depend upon increasing their sales. As we shall see, such companies have enormous leverage with governments and indeed their overpaid executives are frequently seen on the front pages of our newspapers and in our television news advising government on the need for wage restraint, for 'development' and for 'rational' policies. What they mean is that they need such policies to support themselves and the expansion of their companies. The competitive economic system and globalisation of industry and its regulations have ensured that environmentally friendly systems to produce energy cannot get off the ground. They cannot obtain subsidies and even if they could, the influence of existing enterprises – or those with vested interests in them – ensures that they will not receive adequate political or government support. Thus under our present system we are faced with the continuation of the use of oil, gas and coal which produce incredibly expensive energy instead of the alternative energy sources which, if considered in a 20-, 30- or 50-year timeframe, are many times cheaper because they cause no environmental effects. This example alone illustrates the impotence of our present democratic system in matters of international significance. Hope for the future may lie in the early exhaustion of world-wide oil fields. At present the discovery of new oil fields is far behind production and some experts believe that most of the major fields have been discovered. In which case prices may rise considerably and renewable energy systems will be developed.

How much energy do we need to be healthy? Not much. A study in Adelaide has shown that there is a minimum level of energy below which health declines rapidly but above which there is no marked gain in health. Yet the world situation regarding the consumption of energy is frightening. Western countries such as the United States and Australia consume 100 times more energy per person than do poor countries in Africa and Asia. These poor countries want Western-type development and this 'development' will mean an increase in their energy requirements. Yet for the whole world to consume energy at the same rate as the United States or Australia, five times more energy would be required. This would be unsustainable in terms of its effects on global warming. Nevertheless we should realise that this degree of consumption – five times the present rate – *will have arisen by the year* 2030 if we continue as we are, even without correcting the disparities between rich and poor countries. All these forecasts were available for the conference in Berlin (March 1995) when countries such as Australia and the United States, with heavy lobbying from their major companies, failed to deliver any perceptible change in their policies. Since then there has been a move to quotas for the production of greenhouse emissions, supported particularly by European counties, but one wealthy country, Australia, a big producer of coal, oil and gas, has opposed this.

There are ways in which these forecast increases could be tempered, in particular by constructing buildings that save energy and by converting carbon dioxide into other substances in power stations. Again, such actions require finance and for political reasons this cost is not allowed to fall on the producer and consumer; this is the main obstacle to providing any solution. Research is also beginning to indicate that vegetation and soils can trap large amounts of carbon dioxide. In simple terms, rainforests are being destroyed at the rate of one to two football fields per second and these forests alleviate greenhouse effects. The vegetation of grasslands is also important because the introduction of deep-rooted grasses can fix large amounts of carbon dioxide into deep layers of the soil. The maintenance of the world's vegetation cover is essential.

Modernise at any cost: a case of folly

There is a serious lack of insight into energy issues at a political and corporate level. Worldwide there are hundreds of cases of folly. Top of the folly list must be the Far West Electrification Scheme in New South Wales, Australia. In this scheme, 4200 kilometres of wire were put across one of the most remote parts of Australia to provide power to the two small towns of White Cliffs and Tibooburra so that their inhabitants and about 190 homesteads in the vicinity would no longer have to use their local generators. At the time solar power was already fulfilling many of the local needs and there was financial assistance for people to establish solar power units. The final cost of providing electrification was 32 million dollars, whereas local solar systems or local generators could have been provided for a total cost of about nine million dollars. The 32-million-dollar cost takes no account of the huge loss of electricity through the extensive lines themselves – and this is significant, given the resistance of the wire and the distance travelled. This incredible story is documented by Gavin Gilchrist in *The Big Switch* and illustrates the importance of pork-barrelling, political opportunism and official incompetence in the management of environmental issues.[3] Equally appalling examples can be found all over the world.

Consequences of fossil fuel use

The consequences of burning fossil fuels are many and some are outside the scope of this book on health and ecology. In health terms, pollution of cities by car engines is responsible for considerable ill-health particularly in relation to the respiratory system. And there is growing suspicion that the great increase in asthma cases, particularly in urban centres, is due to this pollution. These diseases are not ecological. The two main ecological consequences of the burning of fossil fuel are acid rain and global warming. The former has already happened and will continue, although it may be alleviated by the use of filters on power stations. Global warming has commenced and its major effects will be increasingly felt and may be significant in just 50 years. The health effects of these ecological disasters will now be considered.

The formation of acid rain is discussed in Chapter 1. There are several potential dangers to human health in regions where acid rain falls. These relate to the leaching of heavy metals into drinking water, affecting humans directly and also animals in the food chain such as fish. However, the main effect on humans is to reduce the yield of crops by acidifying soil and damaging their foliage directly. This mechanism is part of the pollution-induced reduction in crop yields in the northern hemisphere over the last few decades.

Global warming is a monumental and globally threatening problem for human health. The physical consequences of global warming – heat stroke, inundation, natural disasters, falling food production – are discussed on page 86. All will affect our health and well-being. But we also face unimagined changes in disease because of the spread of infections.

Most infectious diseases depend for their spread on local environmental conditions. The common cold and influenza viruses, which are transmitted by coughing and sneezing, spread much more easily in crowded conditions. Crowding in poor countries increases the chance of contamination of drinking water by cholera. But many infections depend on temperature, moisture and other species for their transmission to humans. A species that carries or transmits an infection is called a 'vector'. The mosquito is the vector for malaria. Malaria will spread greatly with global warming. It will be only one of many infections that spread but it provides us with a good example of ecological change providing the conditions necessary for the spread.

Malaria and global warming

Malaria is due to a protozoan parasite called *Plasmodium*. A protozoan is an organism of one cell which can live in moist conditions. The *Plasmodium* which causes malaria has a long association with apes and man in the Old World (Europe, Asia and Africa) going back millions of years. The disease is transmitted via a vector, the mosquito, so that when it bites the human the malarial parasites are regurgitated from the mosquito into the human tissues. The mosquito itself becomes infected by biting and swallowing blood from an infected human. Once in the human body the malarial parasites multiply, destroy blood cells and can also invade other parts of the body. The infected person becomes anaemic (short of blood), debilitated and susceptible to other illnesses. Like Jason in Irian Jaya, a person with malaria usually has recurrent high fevers, together with headache, nausea and vomiting. There is then profound sweating which leads to a fall in temperature and the patient recovers from this attack, only to have another after several days. There are different species of malaria, the most feared being *Plasmodium falciparum* which has developed resistance to most drugs used to treat malaria and can lead to severe complications in vital organs. A high proportion of the red cells of the blood can be destroyed which leads to complications in many organs such as the bowel and the brain. The person with malaria of the brain (cerebral malaria) suffers from headaches, fits and then coma and death.

The presence of malaria depends upon the environmental conditions of temperature, rainfall and humidity. For this reason, in one small country, Papua New Guinea, malaria is a scourge of the lowlands and the tropical river valleys but is uncommon in the highlands because the cooler physical conditions are unsuitable for the survival of mosquitoes. The mosquitoes that transmit malaria can only survive where the average temperature in winter is 15 degrees

Celsius or above and optimal conditions are between 20 and 30 degrees Celsius with a humidity of at least 60 per cent. The worldwide distribution of malaria is totally dependent on these factors. Therefore as global warming occurs malaria will spread to areas that have not experienced it before or have not experienced it for centuries. There is firm evidence that that spread has commenced, for in East Africa, Papua New Guinea and Madagascar, where malaria was previously confined to the lowlands, it has now moved into the highlands. It is interesting to note that malaria was endemic in Europe many centuries ago – indeed the word malaria comes from the Italian *mala* and *aria* meaning bad air. Malaria was endemic in Greece in the fourth century and precise descriptions of the fevers were given by Hippocrates, the father of modern medicine. We have to assume that the environmental conditions at that time were favourable and in particular that the widespread marshlands in many coastal areas of the Mediterranean Sea encouraged the breeding of mosquitoes.

Computer modelling suggests that malaria will return to much of Europe, including the British Isles, as global warming occurs. It will gain a hold in the eastern United States and Canada and in Australia it will spread to parts of South Australia, Queensland, New South Wales and Victoria.

The control of malaria

Malaria is a disease that fills the medical scientist with dread. Despite extensive research it has not been possible to find a suitable and effective vaccine. In fact the reader should think of malaria as similar to influenza in that influenza changes its condition on a regular basis so that new vaccines are required as new strains of influenza emerge. Malaria is able to evade the human immune system and furthermore the *falciparum* form seems capable of becoming rapidly resistant to each new drug. The initial attempts to control malaria worldwide were based upon destruction of mosquitoes in swamps, lakes, watercourses and forests. The chemical DDT was used widely around the world and was extremely successful. However, the malarial parasite has now become resistant to DDT and furthermore DDT remains in the environment, causing major ecological damage to other species.

In conclusion, malaria illustrates vividly how changes in our environment and in the ecology of the mosquito will lead to a spread of this dangerous disease. All the evidence suggests that the control will be costly, requiring billions of dollars. The fact that research has not yet found a solution also reflects the fact that it is a Third World disease at the moment and not worthy of the attention of the big pharmaceutical companies for those with the disease would not be able to afford the cure. When we look at all the other diseases that will spread with global warming, affecting humans, plants and animals, we must ask ourselves why we are prepared to allow global warming to occur, for in the long run it would be cheaper to have an immediate revolution in our energy producing systems.

A plague of parasites

A parasite lives off another species of animal or plant. Malaria is a parasite that will benefit and spread with global warming but there are many more.

Anwar was 31 when he first vomited blood. It was a frightening experience and one that left him faint and weak. He knew of others in his village who had had the same problem and had not lived very long afterwards. Anwar was one of a family of ten children living in a remote part of the Nile Delta. For generations his family had survived on subsistence farming of rice and animals. They lived in a small community surrounded by rice fields interspersed with a few pastures. Children and workers were accustomed to the soggy or wet fields as they worked on their crops. Two days after the first incident, Anwar vomited blood again but this time he lost consciousness and was transported by his family to a large sprawling hospital in Alexandria where he received a blood transfusion. His liver was found to be greatly enlarged because of infection with parasites. He was commenced on drugs to kill the parasites but his liver was irretrievably damaged. The pressure of blood in the veins around his liver had increased so much that they were bulging into his stomach and had burst, causing him to vomit blood. The outlook was that he would vomit blood again, almost certainly fatally. Only a liver transplant would save Anwar's life and what hope was there of that in a developing country?

As a child, Anwar had been infected by a worm called *Schistosoma* (bilharzia) by walking on infected ground with bare feet. The tiny worms had penetrated the hard skin of his feet, burrowed into his blood vessels, passing through his lungs and the walls of blood vessels. The adult worms had taken up residence in the large blood vessels which supply the liver and in the liver itself. Over many years his liver was gradually destroyed.

Like many other sufferers of schistosomiasis Anwar had passed the infection to others. When working in the paddy fields it was convenient to defecate under some trees at the edge of the field. The eggs of the worms were excreted in his faeces and when these faeces were washed into the water of the paddy field by rain, the eggs hatched and the worms penetrated a water snail. There they matured and restarted the cycle of infection when they were released into the water and penetrated the feet of other humans. The presence of the worms in the human body causes disease of the liver, bladder and kidneys; a common cause of death is from failure of the liver or vomiting blood. The inflammation set up in the bladder is a cause of bladder cancer and indeed this is the commonest cancer recorded in Egypt where schistosomiasis is a common disease.

Six hundred million persons are exposed to schistosomiasis worldwide and 200 million of these are infected. The disease occurs in Africa, South Asia, the Caribbean and South America. A different form of the worm exists in China, the Philippines and South-East Asia. Obviously the problem will be severe in wet countryside with irrigation channels and rice farming, where people tend to work in wet fields with bare feet. The cycle of infection is also dependent upon a particular snail which carries the young worms and again we see the dependence of this disease on changes in climate and in particular global warming; it is yet another ecological disease. The snails are only infective during the warm summer months. Thus as global warming increases, they will be infective in Egypt throughout the whole year and will become infective in other areas as these increase in temperature. The areas to which they will move will be unpredictable because of the dependence upon changes in rainfall as well as temperature. However, it is certain that they will spread north and south from the present tropics and sub-tropics. Schistosomiasis is a treatable condition and the cost of drugs is not very expensive. However, its prevention requires expensive measures involving improved sanitation, community education and control of the fresh-water snails involved. Quite simply it will be another of the costs that has to be balanced against our neglect of global warming.

We have selected two parasitic diseases, malaria and schistosomiasis, to describe likely events once global warming occurs. There are many others that will spread. Some like malaria are transmitted by the mosquito. One of the most feared is filariasis. In the tropics there are 900 million people at risk of this disease and one out of ten of these becomes infected. When the mosquito bites it transmits to the human, the infant stage of a worm which grows and mates in the lymphatic vessels of the human body. Millions of tiny worms are released and circulate at night in the bloodstream so that when the person is bitten by a mosquito this mosquito takes up some of the tiny worms and they are transferred to other humans. As a result of the infection the lymphatic vessels in the limbs become blocked and the condition of elephantiasis results. A leg may increase its circumference several times and become grotesque. As a reaction to the worms the patient has fever, pain and inflammation in the affected areas of the body. The treatment is with drugs to eradicate the worms but the cosmetic effects of the condition are difficult to treat.

A plague of viruses

Most of us, when we think about viruses, will think of influenza or the common cold, or the need to vaccinate children against mumps, measles and whooping cough. Travellers may need to be vaccinated against exotic-sounding diseases such as yellow fever and Japanese encephalitis. Above all, the entire community now recognises HIV/AIDS as a viral disease. But infection with HIV does not fit the normal pattern of transmission of viruses. HIV is transmitted sexually and by blood; it does not have to survive in the environment so is not beholden to rats, mice, mosquitoes, lice and the many other living things that transmit viruses to us. There are dozens of viruses in the tropics which inflict serious illness on humans.

Dengue is a disease of tropical and sub-tropical countries, especially their coastal areas which are heavily infested with mosquitoes. The infection commences with a severe fever lasting for over a week, headaches and a rash. More severe forms lead to uncontrollable bleeding and death. There is no effective treatment. Dengue is already an occasional problem in north-east Australia. It has recently increased in incidence in the Americas and will be a prime example of a viral disease spreading because of the changing ecological conditions of global warming. Japanese encephalitis, caused by a virus present in pigs and water birds, is also transmitted to humans by mosquitoes. It causes thousands of deaths each year in Asia and is a major problem in China. The disease has been recorded recently in Indonesia and Papua New Guinea. It seems inevitable that it will spread to northern Australia but significant spread will be enhanced by global warming. Vaccines are used on hundreds of millions of people in China but Australia would require a new vaccine produced under local health regulations.

Yellow fever is caused by a virus which infects the monkeys of the tropical rainforests of western Africa and South-Central America. It is transmitted by mosquitoes when the monkeys are near to man or when man invades and fells the tropical forest. Once the virus is prevalent in the community it may be transmitted from person to person. Yellow fever causes a severe fever with headaches and backache. There is nausea and vomiting and eventual jaundice (the person turns yellow). Death is from failure of the liver and from bleeding with the vomiting of blood. Vaccination is available and lasts for approximately ten years. Yellow fever is expected to increase as the conditions for the proliferation of mosquitoes improve in areas with increased rainfall and temperature. Kenya has experienced its first epidemic of yellow fever in 25 years. There is concern that the virus will spread into Asia, introduced by international travel. The World Health Organisation has contingency plans for such a happening.

Global warming will not only bring about the spread of tropical viral diseases, but viral diseases already existing in more temperate countries will widen their hold within these countries.

Trevor, aged 45, worked in Perth, Australia as a government adviser and lived at the edge of the suburbs. His job was stressful and his leisure time scarce. On one day off, intended for golf, he took to his bed with headache and fever. Once before he had had influenza and the symptoms seemed just the same this time, including the aching in his joints and muscles. Three days later the fever and headaches had gone but his joints were so sore he could scarcely walk for the pain. His doctor was called, blood tests were done and his extremely painful knees, ankles and shoulders were examined. But there was nothing to see. In arthritis, the joints are not only painful, they look swollen and are often hot. The doctor suspected stress as the cause, especially given Trevor's occupation. Painkillers helped only a little and Trevor languished at home, tired, debilitated and in severe pain. He was saved from a diagnostic label of 'stress' or 'work-phobia' by the result of the blood test, which showed he had Ross River fever. Patients with this disease say that it devastates their lives. Because their joints often look normal they cannot cope with the possibility that others may think they are malingering. Trevor had six months off work and still has sore joints, although they are much better than previously. There is no cure.

This mosquito-transmitted virus is increasing in Australia: there were 6000 people infected in 1996. The increase is most common in coastal regions with salt marshes which encourage the mosquitoes. But it also occurs in arid regions, emerging after floods which lead to desiccated mosquito eggs hatching into infected mosquitoes. Global warming will enhance the spread of Ross River fever as the conditions for mosquitoes become more favourable.

Cholera

At present, diarrhoeal infections cause three million childhood deaths each year. In Chapter 5, the outbreak of cholera in England in the 1830s is described; it was a disease that thrived on poverty, poor hygiene and the contamination of water supplies with faeces. Cholera is still active in many parts of the world. Until recently it was believed that cholera could survive in water for only one week. It is now known to survive much longer in marine ecosystems that are disturbed with the pollutants of urban living. This polluted water supports the growth of phytoplankton (algae) in which the cholera bacteria survive and emerge when the water warms. There is now scientific evidence that cholera will increase not only because of poverty, poor hygiene and overpopulation but because of urban pollution and global warming.

This chapter has described a few health and disease scenarios of global warming. To do justice to the problem we would have to write a medical textbook. And it is worth remembering that it is not only humans who will experience a new exposure to disease. The viruses, parasites and bacteria that afflict plants and animals are already on the move and will threaten our livestock, cereals, fruits and vegetables.

Global warming, so far, has been the subject of much talk but little action. Some of the wealthiest countries such as Australia have prevaricated most, possibly because their governments are more susceptible to lobbying by management of big international companies. Despite the evidence on global warming to date, those who feel their interests are at stake continue to argue that more scientific studies need to be done before drastic action is taken. At a conference on climate change in Geneva, Switzerland in 1996, the governments of 134 nations committed industrialised nations to legally-binding targets for reductions in greenhouse gases. This is a first small step for humankind. The government of Australia, a rich country, did not support this step; reportedly it was extensively lobbied by the coal industry.

---

1. A.J. McMichael, A. Haines, R. Slooff, S. Kovats, eds., *Climate Change and Human Health* (World Health Organisation, Geneva, 1996).

2. C. Savage, columnist, *The Australian*, 15 April 1995.

3. G. Gilchrist, *The Big Switch* (Allen & Unwin, St Leonards, New South Wales, 1994).

CHAPTER 5

A Plague of People: Population, Food Supply and Health

Most individuals who are prepared to look at the evidence now believe that the rate of increase of the world's population is a prescription for potential catastrophe within the next 100 years. The increase in population is doing irreversible environmental damage which brings closer and closer the day when there will be no more land to utilise and starvation will ensue. There remain those who dismiss this impending calamity for ideological or religious grounds, but the world's scientists and other thinkers now agree that there is such a threat. These issues will be discussed in this chapter and in particular we will explore their effect on the ecology of the world and on our future health and well-being.

In the mid-1950s in China, Chairman Mao galvanised the entire population into the 'Great Leap Forward' – an ideologically driven plan to greatly increase productivity. However, it came to be a great leap backwards for the environment of China and a leap into oblivion for many of its inhabitants. Every scrap of household metal was used to increase steel production, and forests and orchards were chopped down to provide fuel for furnaces. The dictates on agricultural practices included 'deep ploughing' of fields, which led to the burying of top-soil. Food production collapsed and catastrophic famine set in. Throughout the history of the human race there have been episodes of famine. In all famines, people behave in the same way for they are driven by the strongest instinct of all – survival by eating and drinking. They eat grass, weeds, the bark from trees, the straw from huts, the leather from boots, mice, cockroaches, beetles from cattle dung and the worms from human stools. The countryside is laid bare of living things and suffers massive environmental and ecological damage.

The purpose of telling the story of this recent famine in China is to point out that although more than 30 million persons died of starvation, this great loss of life scarcely made a blip on the rising curve of China's population, which has now reached 1.2 billion.

The world's population explosion is a serious threat to health. This explosion will lead to the calamity of displaced populations, social conflict and warfare, rapid environmental degradation and starvation. We already know from our television screens what these events mean for health and well-being. Distressingly, we can see little, apart from vast epidemics of infectious diseases, that could stop the world's population doubling again in the next 30–40 years. As an example of the problem we will look in detail at China, which has had success in cutting its birthrate by dictate (a dictate that cannot be made in most democratic countries for it would not be accepted) yet calamity is still likely in China because of the weight of population, scarcity of arable land and the fall in food production. If China fails, what will happen to the many countries which retain a much higher birthrate?

The predictions and how we view them

In 1993 the scientists of the world issued dire warnings about the world's increasing population. The first was from two of the world's leading scientific organisations, the United States National Academy of Sciences and the British Royal Society. They stated that if our present activities continue unchanged, science and technology may not be able to prevent irreversible degradation of the environment or continued poverty for much of the world. They believed that the future of our planet was hanging in the balance due to the expansion of the human population. The second was the 'World Scientists' warning to humanity. The statement was signed by 1680 scientific leaders from 70 countries and included 104 Nobel Prize winners. They stated that humans and nature are on a collision course and that we are fast approaching many of the Earth's limits. Pressures from population growth put demands on nature which cannot continue and we have only a few decades to change our ways.

There is a fundamental problem with the structure of human society and our way of thinking. When such statements are issued, the majority of those who could conceivably act upon them, don't hear them. The media, whose owners have a mind-set of 'growth' which favours their own enterprises, continue their usual diet of showbiz, murder and sport. This diet remained unchanged on most commercial television channels on the days these international statements were issued. The fundamental difficulty is that most of the communication systems in the West are owned for private gain. In the opinion of the authors, only the medical press responded to these statements from eminent scientists with appropriate concern and debate. In fact, the message had a deep impact within sections of the medical profession. But even when the message is received, it is not necessarily acted upon. We don't want to know, we are overwhelmed. If we cannot influence our governments to stop wood-chipping our remaining forests and building new motorways in already polluted cities, how can we get them listen to our concerns on world population?

The increase in the world's population each year is likely to remain above 86 million until the year 2015. In 2015 the world's population will be between 7.1 billion and 7.8 billion, and in 2050

between 7.9 billion and 11.9 billion. Hopefully the reader's mind
is not numbed by these numbers for there is more bad news to
come. The United Nations Population Institute announced on
28 December 1995 that the world's population had increased by
100 million people, the biggest 12-month increase yet. If current
rates continue the world's population will reach 14 billion by 2050.
Three billion young people will enter their reproductive years in the
coming generation. Ninety per cent of this population growth is in
poor countries. Whilst 30 countries have reported declining birth
rates – including Thailand, Kenya, Zimbabwe, Indonesia, Mexico
and Brazil – 80 countries across the globe are reproducing at a rate
that will see their populations double within the next 30 years.
Forty-three of these countries are in Africa alone.

Will economic growth solve the problem?

Despite these figures some people remain optimistic about the
human population problem. They point to China as an example of
what can be done if there is the political will. In 1965 the average
Chinese woman had 6.5 live babies in her lifetime, but by 1996 this
figure had fallen to 1.4. However, as we will explain later in this
chapter, China is a special case and may already have passed the
point of no return – its population cannot be sustained even at
current levels.

The optimists believe that what is known as a *demographic
transition* will occur and will stabilise the world's population. This
means that with increased economic growth, prosperity, moderni-
sation and a decline in child deaths, a shift to small family sizes will
occur. They believe that it is poverty and lack of development that
causes overpopulation. This is correct to a certain degree. It seems at
first glance to describe fairly accurately the population situation of

the West, where in recent times birth rates have fallen because of prosperity. However, others such as Virginia Abernethy disagree.[1] They believe that expanding economic opportunities are followed by rising fertility, and worsening conditions are followed by declining fertility. The United States and Western Europe experienced baby booms after the Second World War during periods of economic expansion, but when harder times came, the size of families contracted. In fact this explains the growth of human population over long periods. Massive spurts in the population growth rate have occurred in times of major technological change – such as the beginning of the agricultural revolution and the industrial revolution. The most astonishing growth of human population has been in the past 40 years – more people have been added to the world in this period than in the previous three million years. The reason for this is that public health measures have lowered Third World death rates without there being a lowering of birth rates. The world's population is doubling every 40 years.

What turns the problem of population growth into a crisis is that it is occurring at the same time as a massive destruction of the environment and ecosystems worldwide. The population problem cannot be solved by programmes of massive economic growth for such massive economic growth inevitably leads to environmental destruction.

A study of China

We can illustrate this dilemma by considering the case of China – regarded a success story by the optimists. Some predict it to be the world's next economic superpower. Yet a report released by Beijing's Ministry of Agriculture and a Japanese aid agency in 1995 predicted that China would become increasingly dependent upon food imports, particularly grain, because of its expanding population and a diminishing supply of arable land. The report noted that 'if China has to import a large quantity of grain ... the impact on the international grain market will be incalculable.' China, with 1.2 billion people, has 22 per cent of the world's population but only seven per cent of its arable land; yet its population still grows by 14 million people per year.[2]

China's population policy is in conflict with individual freedoms. China's one-child policy was harshly criticised at the United Nations Conference on Women held in Beijing in 1995, not just because of forced abortions and sterilisations, but because the policy conflicts with Western ideals of democracy and liberalism. So should every Chinese woman be free to have as many children as she pleases?

An article in *The Australian* newspaper on 3 June 1995, entitled 'West's Morals Can't Feed China's Millions', generated some interesting responses. One letter to the editor accused the anti-population lobby of being dominated by 'white First World voices' – the corespondent choosing to ignore that it is the Chinese authorities themselves that are implementing China's one-child policy. Another letter expressed confidence that international trade would feed the masses. These are typical views that fail to recognise the enormous scale of China's population increase.

The magnitude of the problem was illustrated by former Australian Prime Minister Keating, when he said:

China's demand for food ... is growing so fast that its shortage within 15 years could be three to six times Australia's total annual wheat production. Just feeding chickens to satisfy China's demand by 2000 will take more grain than Australia currently produces.[3]

Lester Brown in *Who Will Feed China?* stated that if China was to follow the same trend as Japan in eating seafood then it would require 100 million tonnes of fish a year, which is equal to the entire world catch.[4] Yet as Brown argues the world's catch of fish is decreasing each year.

China in any case cannot pursue the path taken by Japan and Singapore and import the bulk of its food and raw resources. Vaclav Smil in *China's Environmental Crisis* believes that the cost would be too great.[5] The over-exploitation of domestic resources to pay for food would further degrade and pollute the environment. China is already facing a massive environmental crisis created by the pollution produced by its industries and the damage to its ecosystems. Commitments to build huge dams on its rivers to produce hydroelectric power for 'growth' may further reduce food production. Intensive farming and deforestation have weakened its soils. He concludes that there are no technical or economic solutions which can reverse these trends; China, despite its population policy, is already on the slippery slope to a deteriorating future. How can such a viewpoint fit with glowing reports of fast economic growth and consumer spending? As we will see, 'growth' is deceptive, particularly when it has been attained by mortgaging the future – by this we mean that present-day production is based on deteriorating soils, exploitation of forests, use of water which cannot be sustained and pollution of the environment which is not being paid for.

The solution for China? China may have to expand territorially or suffer the internal strife which results from hunger. Indeed it may already be flexing its muscles for it fears 'containment', but the West officially denies it is containing China. No statesman talks about it but all worry about it.

The reason we have discussed China and not India or Indonesia or others is that China has recognised its problem and has a population policy, yet its outlook for avoiding hunger, malnutrition and strife is poor. What, then, is the outlook for many other countries? The population of India in 1990 was 850 million. In the year

2000 it will be 1043 million and in 2025 it is expected to be 2025 million, exceeding that of China.

Our discussion of the problems of China also highlights the conflict between the individual's freedom to have children and the community's need not to have children. In this regard an even greater problem is that posed by some of the world's religions. A leading medical journal, *The Lancet*, has recognised these issues. A 1993 article criticised the Roman Catholic Church and Islam for their opposition to contraception.[6] Even some members of the medical profession with their scientific training could not tolerate this interference with dogma. One letter criticised the article as 'offensive bigotry'. To the non-religious there is incredulity that religious leaders – who by personal example stand for the alleviation of poverty, for justice, for peace and for personal integrity – are being used to sustain a belief that may ultimately destroy these ideals. The difficulties in addressing this issue has led to it being dropped from public debate and action. Somehow this debate must be reopened in a climate of mutual respect

Economic growth in Asia

Economic rationalists see the Asian region as the major driving force of twenty-first century capitalism. They seldom examine the ecological costs of this growth. Let us do so now. At an international food conference held in Canberra, Australia in 1993, predictions were made of mass starvation, social chaos and even war in the region, primarily because of population growth's pressure upon limited resources. The Director-General of the International Rice Research Institute, Dr Klaus Lampe, and other experts, while accepting that major achievements have occurred in food production in the past 30 years, expressed doubts that these achievements could continue given the interaction between population, reduced resources and environmental degradation. The amount of grain produced per person in Asia has fallen for the past nine years as urbanisation expands into farm land. This is forcing farmers into cultivating marginal and mountainous lands, resulting in terrible environmental destruction.

In particular, water is becoming a highly contested resource in Asia. It takes 2000 litres of water just to grow one kilogram of rice. By contrast one kilogram of potatoes requires only 500 litres of water and one kilogram of wheat, 900 litres of water. In Asia water is becoming scarce not only because of the demand for drinking water, but primarily from degradation by silt resulting from erosion, human and animal excreta, industrial and chemical pollution, leached nutrients and often toxic algal blooms. Water shortages in both Pakistan and India have led to India wishing to divert the Indus River which brings water from the Himalayas to both countries. Pakistan has countered by threatening war if this is done. It is possible that a protracted war between India and Pakistan could lead to a nuclear exchange. (According to the CIA, sophisticated delivery systems are not needed for this. The two atomic bombs dropped on Japan did not employ sophisticated delivery systems.) Conflict over water is discussed further on page 121.

How many people can the world carry?

When many people think about the human population problem, they often tend to see it as a problem of sheer numbers of people, perhaps even a problem of how many people in each square kilometre. There are of course places on the Earth where people live very close together – Hong Kong and Mexico City for example. However even in China there are vast spaces where population densities are low – such as in the Gobi desert. Similarly, Australia has areas of low population density, such as its interior desert regions. But it would be a mistake to regard these areas as 'unoccupied'. Much of this land is already *culturally* fully occupied, belonging to Aboriginal inhabitants as traditional tribal lands and it would not sustain additional people. The human population problem is not one of space but of lifestyle and environmental impact. We need to be clear about some basic ecological concepts to understand this.

When we study animals we realise that the carrying capacity of the environment is the number of animals it can support without reducing the environment's ability to support future generations. For humans, carrying capacity is related to the biological quality of life. It is all a question of what lifestyles are adopted and what load or pressure is placed on the environment. Animals can grow in number beyond the capacity of the environment to support them. The number will suddenly fall due to lack of food, to disease or to predators. Humankind in one region can continue to grow by 'future eating', by eating into the ecological 'capital' of the future and in turn causing environmental damage or by importing resources and food, drawing on scarce ecological 'capital' from other regions. A simple example from farming would be, the farmer can maintain the health of his soil by growing only one crop a year, but he might double his income and produce twice as much food by sowing two crops a year – until the soil deteriorates and blows away.

Population and Australia

The concerns of environmental impact can be well illustrated by Australia's increase in population which is generated by both sustained immigration and natural increase. The problem is not caused by massive numbers of people – although Australia has one of the highest rates of population growth of any developed country. The problem, exactly like that of China's environmental crisis, is one of damage to the environment. Tim Flannery in *The Future Eaters* argues that Australia's population of a little over 17 million is already too high and cannot be sustained at this level.[7] Modern lifestyles, even environmentally sensitive ones, may require a population of between six and 12 million if environmental degradation is not to further occur in the oldest, driest and most fragile continent on Earth. To date the immigration policy in Australia has been based upon a perceived necessity for growth. The powerful commerce and business elites have supported increased immigration regardless of environmental consequence. Certain political groups with a racist agenda have used the economic argument to oppose immigration. But both sides are failing to perceive the real threat. An increasing number of experts are coming to see population and immigration as major environmental threats in Australia. The authors believe that some immigration should be supported on a humanitarian, non-racial basis, but not to support Australia's economic needs. We all must realise that Australia is a 'vanishing continent' whose basic ecology is in peril.

Population growth

It is important to understand that human population growth is nearly *exponential*. Every 40 years, one million becomes two, two million four, four becomes eight and so on. Disaster has been avoided by technological innovations in medicine, agriculture and industry, which have also increased at an exponential rate, but as we have seen there is debate about whether this rate can continue. Several writers and scientists have drawn vivid comparisons to bring home to us the true meaning of this increase in population.

James Lovelock in *Gaia* describes the situation as 'the people plague':

> *'Humankind behaves in some way as a bacterium, or*
> *like the cells of a tumour . . . so that, the human species*
> *is now a serious planetary disease.*[8]

A malignant tumour within the human body characteristically grows exponentially. Every few days the number of cells doubles. The effects are not noticed when one cell becomes two or two cells become four, but by the time 20 million becomes 40 and 40 becomes 80 over a few days, there is a lump and then a large lump. In the same way humanity can be seen as 'an ecological cancer', converting all available plant, animal, organic and inorganic matter into either human bodies or into the machinery to support them.[9] We are a malignant eco-tumour, an uncontrolled growth of a single species that threatens the existence of all other species on Earth. Mankind as a cancer on the Earth may be a shocking concept. It challenges our ideals of progress, rationality, science and development. It renders our systems of philosophy and ethics mere dust. For it says that the problem is people.

But we do not believe that acceptance of the Gaia hypothesis necessarily implies hatred of humanity. There is merit in viewing population growth as a destructive ecological process, but we must remember that this model is ultimately a metaphor. However, science has demonstrated to us the enormous impact that population size is having upon ecosystems and it is through science and rationality that we have come to understand this problem. Humans, unlike cancers, can and have acted to limit their growth. We can also choose to adopt ecologically sustainable ways of living so that we do not resemble tumours.

Let us now review some of the foreseeable consequences of population growth upon health.

Food supply

We worry about the growth of population because in the future it will be difficult or impossible to feed this population. However, there are important and powerful voices who do not wish to recognise the problem. To recognise it would contradict their own theories; it would mean that the economic rationalist view of unlimited growth would be wrong. Julian Simon, a University of Maryland economist, is a leading proponent of this view as detailed in *The Ultimate Resource*.[10]

Simon would only recognise the problem of food shortage if the price of food went up. In other words, all arguments have to have an economic basis. His argument is that abundance or scarcity can only be measured by its cost. Since the cost of labour is continuously rising, it is people that are scarce! One does not need to be an economist to recognise, in this world of unemployment, that the rise in the cost of labour results from the inadequate economic system that Professor Simon supports.

Having read all the information they can on population, food production and the environment, the authors didn't know whether to laugh or cry at Julian Simon's statement. As we will discuss in the chapter on economics, the problem is that most economists are unable to debate an issue that goes beyond their own imperfect, unscientific system. This would not be of consequence if so many economists were not in positions of power – in government, multinational companies and politics.

Reading reports from the World Bank and the United Nations it is apparent that world food production, after rising steadily for many centuries, has now peaked and is falling. Cereal grain production has shown a six per cent decline per person over the period 1984–1992. Yet there are still 800 million people malnourished or starving on the face of the Earth. It is estimated that food production will need to double by 2030 if world population continues its present growth. We accept that there are areas where there is an abundance of food, and areas, particularly in Africa, where there are severe shortages. It is thus possible that a more equitable distribution of existing food supplies could alleviate this problem for some time. The maldistribution is related to the economic inequities of our system and there is little political will to resolve them.

The options for increasing food production are discussed by Professor Tony McMichael in *Planetary Overload*.[11] The measured production of any crop depends ultimately on the conversion of carbon dioxide and water into the substance of the plant. To do this, we require adequate water and nutrition for the plant. These can be supplied by irrigation and fertiliser. We can also reduce crop losses with pesticides and we can selectively breed or genetically engineer high yielding strains. Yet another option is increased cropping. The limitations of all these options are now obvious to us: they explain why food production may have reached its peak. Water is a scarce resource in many parts of the world and irrigation may provide only temporary increases in productivity; it may function for decades but is then beset by deterioration in the irrigated soils (for example salination in Australia and other arid countries). Fertilisers do nothing to maintain soil structure and they encourage over-cropping, which leads to a deterioration in soils; moreover, they add nutrients to waterways, leading to the growth of algae which pollute the water. Pesticides also provide a temporary increase in production but they also harm the ecology of soils and pastures, and pests resistant to pesticides soon occur. Indeed the possibility of more plant diseases may create a parlous state similar to that with human diseases and the overuse of antibiotics.

Humanity has become used to the idea that all problems can be fixed by technology. In the past, great improvements in farming techniques and in genetic stock and the use of fertilisers led to rapidly increasing food supplies. Will this happen in the future by means of genetic engineering and by cloning productive livestock? It may. But the use of the precautionary principle should lead us to be sceptical, for there are many factors stacked against a significant increase in food production by this means. The spurt in production due to breeding and genetic engineering may be over already for *whatever the genetics* of a new plant, it depends ultimately for its increased growth on more nutrients and water.

It is accepted that in the world there are some 1.5 billion hectares of readily usable arable land and all is now in use. More land can be provided by more deforestation and the use of highly marginal land. This in itself will have deleterious effects on the environment, with loss of ground water and degradation of existing land. Flooding and erosion will certainly increase, and arable land is further threatened by urbanisation and, in low-lying countries, submersion. Already six million hectares, that is four per cent of the total land available in the world, is lost each year through bad farming practices and degradation.

In addition there are number of general factors causing a decrease in food production. Air pollution in Europe and the United States is thought to be responsible for a ten per cent fall in agricultural yields over the last few decades. Deforestation in large parts of the world has led to flooding and further soil erosion with loss of aquifers because there is insufficient vegetation to retain the water. The UN estimates that in the majority of countries in the world demand for the products of grasslands, forests and fisheries already exceeds sustainable yields. Global warming will cause heat stress to crops, and ozone depletion will lead to further impairment of photosynthesis.

In terms of the health, well-being and cohesion of some communities, intensive agriculture demanded by our global economy has already produced dire consequences. The cohesive subsistence community is replaced by poor seasonal workers for the intensive farms, subject to the pricing and employment vagaries of the competitive world. The sick in many developing countries have become sicker and the poor poorer, malnourished and subject to disease.

Water

A lack of fresh water is likely to be a limiting factor in preventing an increase in world food production. There is information indicating that many countries have already reached the limit of their water supply. This is particularly so in North Africa and the Middle East. The need for water will foster conflict over the great river valleys of the world such as the Tigris, Euphrates, Indus, Nile, Jordan, and the great rivers of China. The dry regions of the Middle East and the countries of South-East Asia will suffer this tension first with severe shortages within 25 years. The need for water is particularly great in rice-growing areas for the growth of this cereal consumes so much; yet these are the areas that feed some of the world's most dense populations. Food production that depends on irrigation is likely to suffer greatly because of salination and soil deterioration, and water tables in many parts of the world are falling because of excessive usage. In developed countries much of the fresh water supply is polluted by industry and farming and the damage to land by these pollutants can be longstanding. Large dams to produce power and water supplies have been found to have a finite life as well as causing environmental problems, yet they are still being constructed in many developing countries.

There is no immediate solution to the water problem. Rich countries making their money from selling oil, which produces the greenhouse effect, can afford desalination plants and there are fantasies to tow icebergs from Antarctica to produce fresh water in Australia! However, there is now acceptance by the scientific community that there are no technological solutions to this problem that could be applied worldwide. Privatisation of water supplies (see page 133) is likely to add to the problems of distribution because profit and not health or equity will be a deciding factor in policy.

Soil

Soil is an ecological community of living things. Its strength is created by bacteria, fungi, roots of plants, decaying organic matter and minerals, all of which depend on each other. A healthy soil has a structure like a sponge, a degraded soil is like sand or dust. The sponge-like structure results from micro-organisms and roots binding together the particles of soil. And many bacteria take nitrogen from the air and allow it to be used for growth by the plants. For a century many of the world's most productive soils have been used intensively for crops. Pesticides have killed many of the beneficial micro-organisms and ploughing has frequently disturbed the soil. The sponge has been breaking down. The process of deterioration is slow but sure and may not be recognised until the soil has been washed or blown away.

In May 1995 a strong wind whipped up a dust storm from the wheat belt of South Australia. The dust reached New Zealand where it stained the snow and ice of the mountains red. It was estimated that 22 million tonnes of top soil blew away, containing nutrients worth 30 million dollars. This occurred because of lack of native vegetation to act as wind-breaks and because of a deteriorating soil structure due to intensive cropping. Worldwide there is significant soil deterioration, a serious problem because it can take hundreds of years for new soil to evolve. Some advanced countries have recognised the need to maintain land in good shape and have established landcare programmes. Overall, however, there is much deterioration that will be beyond repair. Yet we are already using nearly all the arable land on the Earth. There is no reserve.

Fish and the seas

The oceans of the world depict the problem of 'the commons' discussed on page 167. Fish is the main source of protein for a fifth of the world's population. These consumers are mainly inhabitants of the poorer countries who have little or no alternative source of non-vegetable protein. There is little doubt that the annual worldwide catch of fish is falling despite improvements in technology and a trebling of the number of fishing boats in recent years. The reasons are many and are probably irreversible. Many of the world's traditional fishing grounds have collapsed simply from over-fishing, for example the North Atlantic Grand Banks; others, such as the Baltic Sea, have been poisoned by industrial pollution. The Mediterranean Sea is in a parlous condition. It is probable that some species have been fished to extinction. The fishing grounds adjacent to the malnourished populations of Africa, Asia and South America have been greatly depleted, often by the fishing fleets of rich countries. The problem of over-fishing is compounded by many other practices, not just industrial pollution. Coastal 'development' – with destruction of mangroves, run-off from agricultural land, which kills sea grasses, and dredging – all have an impact. The ocean's ecology is also greatly disturbed by drift netting, which destroys many other species. Trawling for prawns represents a marine holocaust, because for every tonne of prawns, ten tonnes of other fish, turtles and corals are killed. Global warming is expected to have unforeseen but probably negative effects on the remaining fish stocks as ocean currents change and the source of fish food, the phytoplankton, are depleted.

There is no better example of man's continued greed and

rape of nature than the recent history of fishing. Fishing is difficult to police so is in effect a free-for-all. Nations threaten each other with gunboats to make sure of their share of the spoils. The poor of the world are being asked to suffer increasing malnutrition in order to supply the fashionable restaurants of Tokyo and New York. The problem of fishing and the oceans epitomises every other ecological problem we have spawned by inequitable distribution, poverty, greed, ignorance, political ineptitude and our dominion over and exploitation of nature.

Starvation

We commenced this discussion by describing the deaths of 30 million Chinese from starvation. At present, deaths from starvation are occurring every minute of the day somewhere on Earth. What does this mean in health terms for the hundreds of thousands affected now and what will it mean to the millions affected in the future if we fail to address the issue? In our hearts we know the answer to this question for the pictures from Africa displayed on our television screens seem to say it all. But do they?

A typical year, somewhere in East Africa. The harvest has failed yet again. The cattle have died, the scanty crops have perished in the fields. Large communities roam through the parched lands in search of food. Ona is two-and-a-half years old and is being carried by her mother; her older brothers and sisters straggle behind. You have seen Ona or dozens like her on television. Her body is the size of a one-year-old, but her face looks old for the skin is drawn tight. Her head seems too large for her puny body. She has ribs that stick out and a blown-up belly. She is soiled for she has diarrhoea. Her skin is cracked and ulcers are forming especially around her soiled bottom, her hair has turned reddish and is sparse. She whimpers, for to cry would use too much energy. At the end of the day she lies and watches, for there is no energy to think or move.

Ona is suffering prolonged starvation. She left her mother's breast too early because the milk dried up – her mother was malnourished. She has not received enough protein to grow, build muscles, feed her skin and bones or to allow the chemistry of the body to function. Her immune system, which should have developed resistance to the many childhood infections, is non-existent, so she has contracted dysentery. The morsels she eats now pass through her without benefit.

The gaggle of people straggle across our television screen and we are left not knowing Ona's fate. But without a miracle to change her circumstances she will die during the next few weeks. Even if we could transport her to a well-equipped hospital she would still have a 20 per cent chance of death. But what of the future of Ona's older brothers and sisters? All of them also suffer from a significant degree of what we call protein-calorie malnutrition. Some will survive, but famine will have marked them for life. The television picture does not show us this future. They will be small, poor physical specimens, unable to play a full part in farming and herding. More importantly, the growth and development of their brains may have been retarded and a reduced intelligence may be their lifetime burden. This is the ultimate in human disease, misery and loss of well-being. Nor does the television picture show the future environmental and ecological legacy which will stunt future generations: the loss of marginal farmland by over-cropping, the ravaging of woodland for fuel, the loss of livestock for breeding, disintegration of the social fabric and the possibility of war over resources.

Infection

In earlier chapters we have discussed the likely resurgence of infections in the coming decades. The responsible factors are the disturbance in global ecology by deforestation and changes in land use, the changes in our personal ecology from the overuse and misuse of antibiotics, and the advent of global warming which will cause existing diseases to spread and to have a greater impact. There will be one other important factor in the increase of diseases due to infection: the sheer number of people and their existence cheek to jowl.

Most infectious diseases are spread by close contact between persons or between persons, insects, parasites and animals. Close contact promotes the chance of catching infection by touching, coughing and sneezing. Crowding promotes those diseases that cause vomiting and diarrhoea by contamination of food and water with germs carried in the faeces. History shows us that crowded rural communities and densely populated urban areas are at risk, as was the case with the Black Death, which killed between a quarter and a half of the population it infected.

One Saturday in October 1831 a sailor, William Sproat of the port of Sunderland in the north of England, had been suffering from a bad bout of what was called 'summer diarrhoea' but he felt well enough to walk to his boat on the River Tees. On his return 'he became very ill, had a severe shivering fit and giddiness, cramps of the stomach, and violent vomiting and purging'. On Sunday he was visited by a doctor who found his pulse was weak, his limbs cold, his eyes sunken, his lips blue and his legs in cramp. He had continuous vomiting and diarrhoea. On Wednesday he died. Within a few days his son also died after suffering the same symptoms.[12]

William Sprout was the first person to suffer from Asiatic cholera in the British Isles. Presumably the infection was brought in infected water on his ship. His diarrhoea, which would have contained millions of cholera germs, infected his close relatives and spread to local water supplies, which were in those days contaminated by sewerage. In 1831 and the subsequent years, the epidemic killed 140,000 people. Such epidemics occur in the world today and are probably increasing. One began in South-East Asia in the 1960s and spread through Africa and has now appeared in Central and South America. As in the British Isles in the 1830s the disease thrives on poor hygiene, contamination of water supplies and poverty. All these conditions exist in many developing countries; as their population increases, cholera and many other infections that cause diarrhoea will also increase, and the implementation of public health measures to prevent them will become more difficult. The spread of cholera will be facilitated by global warming.

These impending environmental and ecological disasters are clearly relevant not just to health, but to our very survival. If the situation is not reversed suffering, starvation and ill-health will be inflicted upon the world's population. The basic building blocks of human well-being – food and water – will steadily diminish, potentially bringing untold consequences in terms of population movement, social disorder and war. It is not our intention to indicate that there is no hope – there is, and we will provide some options in later chapters. Within this chapter, however, the intention has been to paint the present picture and to indicate the likely scenarios if there is no change. All possible measures to protect health and the environment, from recycling to landcare, will become irrelevant if we fail to stem our population growth.

---

1. Virginia Abernethy, *Population Politics* (Plenum Press, New York, 1993).

2. L. Dayton, 'World Starvation Crisis within 10 years', *Sydney Morning Herald* 29 May 1996 p.6

3. Paul Keating, lecture in Singapore delivered in January 1996.

4. Lester Brown, *Who will feed China?* (Norton, New York, 1995).

5. Vaclav Smil, *China's Environmental Crisis* (M.E. Sharpe, New York, 1993).

6. Douwe A.A. Verkuyl, 'Two World Religions and Family Planning', *The Lancet* 21 August 1993, pp. 473–475.

7. Tim Flannery, *The Future Eaters* (Reed Books, Port Melbourne, 1995).

8. James Lovelock, *Gaia: A New Look at Life on Earth* (Oxford University Press, Oxford, 1979).

9. W.M. Hern, 'Why Are There So Many of Us?' *Populations and Environment*, Vol. 12, 1990.

10. Julian Simon, *The Ultimate Resource* (Princeton University Press, Princeton, 1981).

11. Tony McMichael, *Planetary Overload* (Cambridge University Press, Cambridge 1993).

12. R.J. Morris, *Cholera, 1832* (Croom Helm, London, 1976).

CHAPTER 6

BULLS, BEARS AND HEALTH: THE MARKET ECONOMY

The lifestyle benefits of industrialisation are taken for granted by all who have them and are eagerly sought by the citizens of developing countries. In the West we cannot imagine life without the car, telephone, refrigerator, television and computer. But the market system which provided these inventions is changing ever more rapidly. In this chapter we will examine how new economic practices pose problems for our future health and well-being. At a time when the world environment is under threat, when both the US National Academy of Science and the British Royal Society see the future of the planet hanging in the balance, a major economic revolution is occurring. This is the revolution of economic globalisation: the increasingly free movement of money and labour around the world. This revolution is associated with, and justified by, a resurgence of *laissez-faire* or free-market economics. The reader will perhaps be alarmed to learn that the rules and regulations favour the free market over health and environmental considerations.

It is difficult to analyse a civilisation when one is part of it, for we are all influenced to a varying degree by common values. However a cold, analytical and very simplistic view of our society would be likely to come to the conclusion that it is based on money, for there is no organisation without it. Money can be seen as the means whereby millions of humans are increasingly controlled by a few. This power is exerted mainly by multinational companies, global organisations and dictatorships. The companies supply products and services to the masses. Essentially they control supplies of energy, without which industry and commerce could not function; they control communication and information systems, which decide what is given to the people; and they control the money (which is power) through complicated profit-making organisations called banks. War sometimes ensues when control is threatened, especially the control of energy. Individuals may own 'shares' or a tiny part of the company in order to participate in profit (not to make the Earth a better place).

Although each country has rules and laws, the biggest companies are not answerable to these for they can avoid them by moving money and production from one country to another. Each country has a government which may or may not work on behalf of its people but the companies exert their power through these governments, which are either threatened with withdrawal of money, production and jobs, or plied with money to do their bidding. The huge wealth of the companies, particularly the controllers of money, the banks, enables them to form a cabal of international organisations which is autonomous and non-democratic and which makes rules to maintain the power system based on money. In dictatorship, the third control system, a few individuals in one country direct the masses by force and by rules, but these individuals are usually also rich because they work with the companies and institutions. Some countries have a system called democracy, developed to keep the peace by participation and to stop people fighting physically for money, food and power. But the nature of the economic system means that the most essential aspects of life – health, the environment and education for living – are made secondary to many other needs.

Every country is conditioned to aspire to economic 'growth' for this is seen as a way of increasing wealth and jobs. Let us look at the fallacy of 'growth'. 'Growth' measures all items that go through the market place. So unpaid work is not valued. The work of non-government organisations, charities, work in the home, subsistence farming, and voluntary social support systems is not measured. By contrast, smoking, car crashes, burglaries, the financial bonanzas of gambling, crack and heroin sales, the lucrative child sex market and the vast international arms production join with the production of food, clothes and housing to form the 'growth' statistics. (Not all of these activities are legal, but they all contribute to economic growth through their need for equipment and materials, foreign travel, and so forth.) Growth is the shrine before which the majority of political leaders worship. The icon they hold before us is the high annual growth of some Asian countries. These are described as 'tiger' economies. Yet a significant proportion of that growth is being achieved on debt – not debt to the banks, but environmental debt. To achieve this 'growth', forests have been wiped out, seas emptied of fish, soils eroded, and environments heavily polluted by cancerous urban sprawls and the failure to provide infrastructure. These economies are better described as 'rogue elephants' than tigers.

This one economic gauge, 'growth', illustrates more than anything else the absurdity of economic practice. It fails to measure the true economic and social value of a society, it fails to include environmental costs and it fails to distinguish work and production that is of value to society from that which is of no value or is detrimental. Bombs are 'growth' whereas the nurturing of children is economically irrelevant. Recently there has been a move to develop other measures of true progress which take account of social progress.

Some would regard these viewpoints on 'growth' as cynical but many would agree with some of the points that we make. For example, the respected economist J.K. Galbraith has drawn attention to the current revolution of the rich against the poor. We will analyse the flaws in our economic system and indicate how these flaws adversely affect the cornerstones of our ecological health. Today there are hundreds of examples to choose from and tomorrow there will be even more. However, the broad areas we have chosen to examine are water supplies, agriculture, international trade, and mining. All have implications for human and ecological health.

Privatisation and globalisation

Privatisation is in vogue. It is a product of globalisation, which demands that organisations be large to compete internationally – or so we are told. Privatisation is a form of economic globalisation, the surrendering of local assets and resources to the transnational corporations that are nearly always the only buyers of these assets. The philosophy that private enterprise can run services more efficiently is propounded by industry and multinational companies and is accepted by their students and partners, the governments. There are advantages for governments in handing over ownership. They can pay off debts, which they nearly always have, and they can abdicate responsibility for disasters and price rises in sensitive utilities such as water. Privatisation, then, is the idea that government assets, functions and businesses should be sold off or contracted out to private enterprise, which in the free market, under conditions of competition, will do things more efficiently so that consumers benefit by reduced prices and/or an improved quality of service.

This idea has been given a theoretical defence in Australia by the economist Fred Hilmer, director of the University of New South Wales Graduate School of Management, a director of Macquarie Bank and Westfield Holdings and deputy chairman of the brewery giant, Fosters. We shall consider some Australian examples here because in many respects they illustrate the problem of economic rationalist principles better than any other examples that one can find in the world.

Hilmer chaired a committee, along with Mark Rayner, a

director of CRA and Geoffrey Taperell, a partner of the law firm Baker and McKenzie. The inquiry was into international competition policy and resulted in the report 'National Competition Policy' (Report by the Independent Committee of Inquiry, August 1993). The idea here was to divide the public sector into 'business' and 'non-business' activities and then open business activities to competition. This would apply to government boards and departments supplying drinking water and gas, and running railways. Most of these enterprises are owned by the Australian states and make substantial profits, which the states use. National competition policy would see the private corporations who came to run essential services, taxed so that the federal government received a slice of the profits. This scheme was warmly embraced by big business in Australia. BHP's chief executive, John Prescott, said at a meeting in Melbourne of ten of Australia's most powerful industry groups on 21 March 1995 that the Hilmer revolution would help eliminate 'the unnecessary costs of doing business in Australia'. As for these 'unnecessary costs', the (Australian) Industry Commission – a major exponent of economic rationalism in Australia – has estimated that 30 per cent of government services will be contracted out to the private sector, including such basic institutions or services as police, prisons, hospitals and fire protection agencies.

Water privatisation and health

Water privatisation is under way or is being proposed in most countries, developed and less developed, in the world. First World transnational corporations are falling over each other to own or manage the drinking water of less developed Asian and South American countries. This trend is supported by globalist entities such as World Bank. The World Bank uses non-government investment as a condition for approving loans for water development projects. The World Bank also supports privatisation in the First World. It actively pushed for the privatisation of South Australia's water management. Why have the internationalists and globalists moved to snap up water utilities? The reason is, in economic jargon, that consumers have a perfectly inelastic demand curve, meaning that they are not able to alter their level of consumption in response to price rises. Controlling water resources is, in the long term, a licence for printing money.

In the United Kingdom the UK Water Act 1989 enabled water utilities to be privatised. The environmental organisations Friends of the Earth and Greenpeace publicly campaigned against this privatisation, their criticisms being centred upon water quality and the discharge of domestic and industrial wastes into rivers. The European Court of Justice ruled in 1992 that the British government had failed to bring British drinking water up to the European Community standard for nitrates. Friends of the Earth and Greenpeace claimed that the British government shielded the water industry from prosecution for not meeting European community water quality standards. Even so, the big privatised water companies have been prosecuted 157 times for river pollution since privatisation in 1989.

A deterioration in British river quality has occurred. In 1980, 3900 kilometres of rivers were classified as badly polluted; by 1990 it was 4680 kilometres. In 1981, 12,600 water pollution incidents in England and Wales were reported; in 1991 there were 28,143 incidents with only 282 resulting in prosecution. Greenpeace has also noted that 5046 tonnes of toxic pollutants a year flow from factories into the seas through more than 12,000 government-licensed discharge pipes. Greenpeace has also publicised the fact that a government policy statement on the environment has stated that 'the rivers have to be used for waste disposal by industry'. And indeed they are. The British government's own figures, quoted in Greenpeace literature, show that 40 per cent of toxic metals entering the north-east Atlantic came from British rivers, 84 per cent of these toxins being poly-chlorinated biphenyls (PCBs), which affect marine life. The North Sea in particular is in danger of becoming a marine desert because of this pollution, which has greatly reduced fish stocks and resulted in fish deformities.

Research by the University of New South Wales indicates that after the privatisation of water services in the United Kingdom in 1989, domestic water charges had increased on average by 67 per cent by 1995, with increases of up to 122 per cent for sewerage and 108 per cent for water. Disconnection because people could not pay their bills increased by 50 per cent and it is accepted in the UK itself that this has led to hygiene problems and an increase in enteric infection in poor households. The British Medical Association has issued warnings that this situation is unacceptable because of the spread of hepatitis and dysentery. Infrastructural investment levels have not kept pace with the agreed schedules. Companies have passed to consumers the burden of these (inadequate) future invest-ment costs. According to Greenpeace they have also passed on the costs of greatly increased salaries: in the four years since their com-panies left the public sector, the chairmen of ten privatised water companies had salary rises of up to 571 per cent and shared in multi-million pound share and pension packages.[1] Thames Water parti-cipated in these financial bonanzas.

Privatisation of water in Australia is relatively recent but some consequences are evident already. South Australian water and sewerage services were sold to United Water, an Anglo-French con-sortium that included Thames Water, in 1995. The South Australian government trumpeted the financial savings to the public. As became evident, the savings did not result from new expertise for many of the experts were paid off to provide the savings. In 1997, a sickening 'pong' enveloped the entire city as hydrogen sulphides rose from the sewage ponds of the privatised system. Many people experienced nausea. The causes were several but amounted to incompetence resulting in the ponds being overloaded with sewage. An expert report indicated that the monitoring system was inadequate and there was no environmental policy. It was implied that the disaster was partly due to the reduced expenditure.

In many countries, surveys of public opinion have shown strong opposition to the privatisation of water. There is an innate recognition that clean water is essential to life and health. This dates back to one of the greatest advances in environmental health, made by a local surgeon, local clergy and local government officials in London in 1854. In 1853 there were 53,000 cholera deaths in England and Wales. In 1854 John Snow, the London surgeon, conducted two revolutionary experiments.

In the first, he compared the cholera deaths in families receiving water from one private company, Vauxhall, with the cholera deaths in families receiving water from a second company in Lambeth. There were 153 deaths in each 100,000 of the population supplied by the Vauxhall company compared with 26 per 100,000 of the population supplied by Lambeth. Water supplied by the Vauxhall company came from the sewage laden Thames at Battersea Fields. Lambeth drew water from a new site at Thames Ditton. Snow noted that the Vauxhall company water 'was part of that which was past through the kidneys and bowels of two million and a quarter of the inhabitants of London'.

His second study was more courageous. During an explosive outbreak of cholera in Soho he ripped off the handle of the Broad Street pump, thus stopping the community from drinking what he believed to be contaminated water. The epidemic subsided almost immediately.

From 1854 to the present day, communities have wished to retain their right to have a say about the quality of the water they drink. The opposition to chlorination and fluoridation reflects community concern, not a wish to overturn scientific opinion. The public is rarely pacified by political platitudes about the safety of water and it is a wise government and a wise bureaucracy that involve the community at every level of the management of water resources. The cogent arguments against privatisation relate not only to the conflict between safety and the profit motive, but also to community responsibility. Public ownership of a vital resource such as water makes the community responsible. Those in charge are paid by the public and are responsible to the public, not to a narrow spectrum of shareholders. The public sits on the management boards. There is a sense of community in ensuring water resources which extends to voluntary work and surveillance of catchment areas. But the dominant ideology today is that essential services are best managed by harnessing the greed and profit motives of the private sector, rather than by relying on the commitment to serving people that is the ethos of the public sector.

This argument can be understood better by looking at the privatisation of hospitals, which is also occurring worldwide. In Australia the community has played a major role in establishing and running both rural and urban hospitals. The contributions in voluntary time and funds has been massive. Because the costs of health care are so great, governments have contributed financially. At the stroke of a pen, many of these hospitals have been assigned to the private sector. They are now profit-making ventures. The capital built up by community fund raising and time has been given away. But the major concern about these decisions taken by Western economic rationalist governments is that they take away the community's responsibility for its own health – ironically at a time when political philosophy demands individual responsibility for oneself, one's family and one's health. The consequences fall upon Western, individual medicine rather than environmental health, but it is important to recognise them. Because the community's direct financial and voluntary input no longer exists, decisions on health are purely financial and for profit. Surgical procedures that generate profit are promoted; the care of the elderly and mentally ill – which is just as important to the community – does not generate profit so is downgraded. The ability of the community to influence health care for its own benefit has been taken away and given to executives and shareholders. To trace how such decisions come to pass we must look critically at our economic system.

What is wrong with economics

The authors have examined contemporary theories of economics. Our different academic backgrounds allow us to bring different perspectives and thought patterns to our analysis of the subject. What goes on in the minds of economists, we wondered. Why do they say and believe the things they do? We can come to understand the nature of economists by understanding the nature of economics. To do this we need to look at economics as if we were anthropologists or scientists examining the practices of some previously unknown tribe.

Economics purports to be a general theory of human behaviour, tirelessly true and invariant. Individuals, we are assured, are concerned only with maximising their 'expected utility', to obtain the maximum satisfaction of their desires and wants. Furthermore, these rational individuals have infinite wants. They can never be satisfied. Society is nothing more than the sum of the actions of these infinitely selfish little pigs, all fighting to ensure that it is their snout, and theirs alone, that gets into the trough. From all of this, by what is known as the 'invisible hand' – perhaps better called the 'invisible wand' – comes social order. Not only social order, but the best of all possible worlds, provided that the market is fully free to function. So orthodox economists tell us. To believe such a story requires a huge intellectual jump. What the economist believes is rejected not only by common sense, but by most other intellectual disciplines.

However, the influence of economics has been so powerful that its main ideas are now starting to be incorporated into biological theory and social theory through what is known as 'social choice theory'. On the one hand, while some economists tell us that there is no such thing as society, that nothing exists beyond individuals, when they drop their guard, the economy itself is described as 'ailing', 'thriving', 'over-heated' or 'bullish'. It is like a living entity itself, transcending not only human agents but social reality.

Orthodox economists are also – no doubt unintentionally – impostors. Whilst posing as scientists, they offer little more than an exposition of the idea that any productive system works best on the basis of greed. Greed is not merely good – it is great. The more greed to satisfy, the more the rules must be abandoned and the market must be free. So if you do look at economics from the perspective of any other discipline untainted by orthodox economic thought, you will come to realise that it is not based on science, reason, fact or even human need – human need being what all of us need, not what the few need.

To a large extent free market economics follows the needs of the market: the banks their margins, the governments their incomes, the corporations their developments and those really clever ones, their speculative gains. Whatever is left over after 'the market' has fulfilled its needs is allocated to the 'non-essentials' by government budgets. These non-essentials are health, the environment (including primary food production) and education. Some of us would call these 'non-essentials' our 'life support systems' and would construct our economic system around them.

Look at what is happening to 'these life support systems' under this apportionment of monies. They are being used as the necessary reserve funds for the free market to operate; when a financial entity (such as a South Australian bank or Orange County, USA) loses its gamble, the bail-out budget eats into the life support system. To squeeze a health system of funds it is managed by senior bureaucrats, hospital administrators and their political masters, who provide only financial control and take no responsibility for health outcomes. Statements are made that the doctors have full control of decisions to treat or not to treat. In fact, 'full control' means that the profession takes the responsibility to refuse treatment. There is a growing band of 'non-responsible' bureaucrats and administrators to supervise this system, to ensure that falling standards or let us say, standards that are prevented from rising, do not alarm the populace.

Free trade agreements have served as the ideological justification for privatisation and a wind-down of the Canadian welfare state, with 25 billion dollars being slashed from welfare budgets in the past five years. This attitude of cold economic rationalism is well summed up by a statement made in 1995 by the chairperson of the Canadian Manufacturers' Association:

> *All Canadian governments must test all their policies to determine whether or not they reinforce or impede competitiveness. If a policy is anti-competitive, dump it . . . The social programs we've come to depend on . . . we're going to have to abandon. We're going to be shutting down hospitals, like it or lump it.*[8]

Turning to another life-support system, the environment, we find that our future assets are being woodchipped, we cannot provide finance to save ourselves from land salination and we cannot raise the finance needed to stop pollution of our waters and seas by free-market money-making forces. We cannot adequately fund our environmental protection agencies. Our destructive system of financing and interest rates ensures that the custodians of our land have to degrade it or sell it. Yet if you say this to a politician or anyone else inculcated with Western economics, you are seen as a heretic, or mad, or both.

The politicians would perhaps be slightly justified in their support and use of economics if it were an 'objectively true science'. But economics is not, despite its heavy cosmetics of mathematics. The heretical economist Paul Ormerod has pointed out that economists are trained to think of the 'economy' as if this noun referred to something distinct – which it is not.[2] The economy is conditioned by and dependent on the society in which it is found and which is its reason for existence.

In an applied science you decide what you are going to build or make or resolve and you plan your efforts along certain principles to achieve that end. If we have any interest at all in the next generation then we should ask what its life-support systems are and give them priority. We would devise a financial system to support these above all else, with what remains being allocated to the less essential systems. The good news is that there *are* groups of heretics gathering in some universities who think like this. Fortunately our present system does not allow them to be put to the sword for such heresy but their task is formidable. The system that we plan has to be based on new premises, such as the sovereignty loan principle, which could supply finance to life-support systems without interest. The economists of course would be horrified and claim that it would cause inflation, but there is no *scientific* proof that it will or will not, any more than there is proof that inflation is intrinsic to the present economic system. By such means we could provide finance to the custodians of the land, our farmers, to look after their soil, for soils and environments to be repaired, for water catchments to be stabilised. See how many jobs this will produce! Thousands upon thousands. We could allocate money to health according to community expectations. We could eradicate some diseases of developing countries for a *dollar* a head. What would this do for the sustainability of the environment in those countries? Stabilise it in a couple of generations.

What about the developers and speculators? Well, they could play with their own pool of money and interest rates as long as they didn't turn off our life-support system. But indications are that they will, under the present system. At a time when humanity is facing its greatest crisis, when the possibility of the life-support systems of the Earth are in danger of collapsing, we are told by economists that the proposals of ecologists and environmentalists to cut greenhouse gas emissions, to reduce production, consumption and waste production, constitute economic suicide. You can have either jobs or oxygen but not both! To admit that there were major problems with industrial mass societies and our whole way of life would be for economists to admit that economics itself is limited. This would be a blasphemy, something a fanatical priest of a fanatical religion could not admit.

Yet some heretics within the economics profession are prepared to commit such blasphemies. We have already cited one such heretic, Paul Ormerod. He studied economics at both Cambridge and Oxford, was Director of Economics at the Henley Centre for Forecasting from 1982 to 1992, and has held visiting professorships of economics at London and Manchester Universities. He blows the whistle on orthodox economics, showing that the core model of mainstream theoretical economics is seriously flawed. Its vision of an autonomous, rational economic man is an utterly mistaken view of society and economic agents. Ormerod believes that economics is more akin to astronomy and climatology than, say, physics. Economists, like astronomers and climatologists, deal with complex non-linear systems. Small changes in the parts of the system may result in large changes in the whole system, and the whole system is more than the sum of the parts. Contrary to the ideas of orthodox economics, societies do exist as more than merely the sum of individuals. Economic data, like astronomical and climatological data, are incomplete and subject to uncertainty and error. Thus research should involve a careful collection and analysis of data, with theories being built around facts rather than economic reality being twisted and distorted to fit an abstract theory about how economies should be in a mechanical, linear world of equilibrium.

Economic rationalism, with its absurdities of free trade and open borders, exerts the influence it does in the modern world not because of its inherent scientific rationality, but because of ideology. It justifies the existence of global capitalism. It rationalises human greed, selfishness and immorality. If progress is to be made in dealing with the perils described in this book then we must examine the legitimacy of industrial mass societies and their underlying beliefs and philosophies. Economic rationalism is operating as a fanatical religion. However else could it have brainwashed all governments and political parties, have dictated that leaders must generally be economists, have covered the front pages of our papers and television screens with the latest guru pronouncing on interest rates, and have condemned virtually all opposition as heresy? The 'papal bulls' pronounce and the compliant masses supplicate for this care and attention, but the horror of it is that the majority of economists don't give a damn about the basic needs of human beings. Their concern is solely with the growth of the economy.

Famine, food and farming

Malnutrition is a major cause of ill-health. Eight hundred million people on Earth are starving yet 'food mountains' exist in Western countries. We have no structures to ensure equity of distribution yet the situation is clearly recognised by private and public donors. While this may be seen as a major problem, an even greater problem is presented by the fragility of food supplies for all humanity as detailed in Chapter 5. How has globalisation contributed to this problem?

The world is moving rapidly to a free market in agricultural produce. The rules forbid subsidy. The cheapest price for the best product will get the sale – that is the way 'the market' operates. But the cheapest price takes no account of the hidden costs. The greatest hidden cost is the slow degradation of irreplaceable soils under this intense competition. Over the years a host of destructive measures have had to be used for the producers to stay in competition. The farmer might recognise the problem, but has to employ these practices to remain in profit. Add to this the other facets of economic rationalism which beset the farmer – high interest rates, depopulation of rural areas and reduction of services – and the dice is heavily loaded against the future. Unpredictability of pricing for many commodities is another factor. A decision to grow wheat and produce a good crop can be made on the basis of an international price of 250 dollars per tonne. By the time the wheat has grown, the price is 170 dollars. This occurred in Australia in 1996. Such fluctuations provide little chance of budgeting for care of infrastructure – the soil.

If we were to design an agricultural structure that would protect the environment and its ecology we would look at the cost of repairing and sustaining the soils, and the cost of sustainability for the farmer, the farmer's family and the supporting community. The funding of agriculture would be removed from 'rational economics' so that finance was long-term and low-interest. These principles of costing should be used in support of all life-support systems.

Mining, ecology and health

Another sorry example of the sort of world the twenty-first century looks like becoming is supplied by the Ok Tedi mining venture in Papua New Guinea. Ralph Nader and many others view this as 'the biggest single mining devastation of the environment in the world'.[3] Indeed, the Melbourne law firm Slater and Gordon in 1994 filed a four-billion-dollar compensation claim on behalf of the Fly and

Ok Tedi River communities, against BHP (a 60 per cent owner of the mine) in the Victorian Supreme Court. The Papua New Guinean landowners see their environment as having been destroyed by the waste that has been dumped into the Ok Tedi River at the rate of 80,000 tonnes a day. There is no question in this dispute that significant environmental damage has been done. A compensation package of 120 million dollars has been prepared for the Fly and Ok Tedi River communities by BHP and the PNG government drafted legislation called the Eighth Supplemental Agreement to make other general compensation claims a criminal offence. Plaintiffs were turned into criminals at the stroke of a computer key. The draft legislation had been prepared by BHP lawyers. BHP was then found to be in contempt of court by the Victorian Supreme Court, but later won its appeal against that decision on a technicality.[4]

BHP's point of view has been expressed by its General Manager, P.J. Lavers.[5] BHP claims that it was too costly to build a tailing retention system in an earthquake-prone area receiving ten metres of rainfall per annum. So the only option, in their opinion, was the controlled discharge of tailings. The company recognises that it has not applied the same standards that it would apply in say, Australia, but argues that PNG conditions are not Australian conditions.[6]

Having adopted a desire to follow Western paths of development, PNG has locked itself into this situation. Journalist Peter Fries, in a commentary on this situation, sums up how:

> With a crumbling economy, foreign exchange reserves depleted (20 per cent of which came from Ok Tedi), the World Bank and the International Monetary Fund on its back, and civil unrest, any change to the revenue stream from the mine would be fatal for PNG.[7]

BHP argues with the impeccable logic of the economic rationalist that all of their actions do make sense from the perspective of profit maximisation. They have done nothing wrong – they are merely following the inner logic of a transnational corporation. They are good economic agents. They are profitable. In terms of the health and preservation of human communities and the environment, however, their actions are appalling.

The authors believe that this situation is unacceptable in terms of the future of human kind. A significant area of PNG, inhabited by indigenous peoples who obtained a sustainable living from river fishing and riverside gardens, has been poisoned for an indeterminate period, perhaps hundreds or even thousands of years. There is evidence that the ecological effects extend into the surrounding coastal waters. Who is to say that the loss of this fertile land for hundreds of years won't cause an economic loss which is greater than the value of the mine? The human loss and suffering is also unacceptable to any civilised society. This type of mining, entailing loss of land used by indigenous peoples, also occurs with the Porgera mine in PNG and with the Freeport mine in neighbouring Irian Jaya. The mining companies and the governments of these countries point to the profits and the education and health facilities provided to the local populations around the mines. We must ask ourselves whether these are an appropriate trade-off for the long term environmental, ecological and health losses described above. Under an economic system which would take heed of the future, the cost of a tailings dam would have to be included in the cost of production. On this basis the viability of the mine would be decided. BHP shareholders – us (for we all compulsorily invest through our superannuation) – have grown fat on the misery of indigenous peoples and generations to come.

The perils of international trade

Which organisations are deciding our future? GATT (General Agreement on Tariffs and Trade) is an organisation which, since it came into operation in January 1948, has promoted the expansion of international trade through the removal of tariffs (tariffs being taxes placed upon goods brought into a country) and other restrictions on international commerce. GATT operates together with the International Monetary Fund, the World Bank and various free trade blocs and associations – such as APEC (Asia-Pacific Economic Cooperation Conference), ASEAN (Association of South-East Asian Nations), EC (European Community), NAFTA (North American Free Trade Agreement) and the proposed Free Trade Area of the Americas (FTAA) – to produce a world where transnational corporations are free to roam the world in search of the labour and environments that offer maximum profitability for their enterprises. All of these organisations and agreements exist to ensure that markets are deregulated and that nations are open to the penetration of the transnational corporations.

The IMF and World Bank are not democratic. Primarily they serve the interests of investment capital. In developing countries they have sought outcomes that foster the operation of foreign traders and investors. In a globalised world the World Trade Organisation's (WTO) 'dispute settlement panels', composed of WTO experts, will have the final authority to overturn national laws and policies if they violate WTO rules. Both NAFTA and APEC will have the power to review state and federal laws in the United States and Australia to ensure that they do not pose any barrier to free trade. There must be no discrimination against foreign investment and no preference arrangements to assist local industries and employment. Subsidies for the development of renewable energy, by definition based upon local resources, are forbidden. How can this be reasonable when the development of renewable energy is essential to counteract the greenhouse effect, which is probably the greatest threat to our health and future well-being?

Already there are countless examples of free trade rules being detrimental to health, some of which we have discussed in this Chapter. Corporations have used NAFTA to overcome controls on the introduction of genetically engineered growth hormones in dairy cattle, a range of recycling measures, food safety regulations and conservation programs in Canada (US companies can take legal action against many Canadian social and environmental policies on the grounds that they are unfair competition, restricting free trade). Even restrictions on tobacco advertising have been challenged. But economic globalisation has not resulted in limitations on the lead content of wine or in 'gas guzzler' taxes to create incentives for people to buy fuel-efficient cars. The allowable level of pesticide residue in Australian food imports can also be challenged under GATT and if found to be 'too strict' would constitute a non-tariff restrictive trade practice. Consequently, Australians must accept imported fruit with higher pesticide levels than deemed satisfactory by the Australian parliament.[8] Pesticides and some agricultural chemicals are implicated in the development of lymphomas, cancers of the organs which produce blood. In 1992 the United States passed a Marine Mammal Protection Act. This Act stipulated that tuna caught in nets which destroy dolphins could not be imported into the United States. Mexico challenged this before the GATT Dispute Resolution Panel on the grounds that it was a non-tariff trade barrier. Mexico won. The European Union is set to challenge a range of US federal and state laws relating to environmental and safety standards, claiming they are non-tariff trade barriers.

One would expect the economic merits of economic globalisation and free trade to be overwhelming given the power that global organisations have usurped. However, there is an environmental and social case against free trade that we regard as decisive.

This is particularly so with free trade that enhances greenhouse gases. More than half of all international trade involves the simultaneous import and export of essentially the same goods.[9] Thus we have the absurdity, nay the economic treason, of Australia, for example, being flooded with Brazilian orange concentrate while 3000 Australian farmers who could supply fresh local fruit crash into bankruptcy. The Brazilian citrus industry does not consist of native farmers making good: it is largely transnationally owned fruit corporations who have planted orchards after extensive clearing of rainforests.[10]

In *The Myth of Free Trade* Ravi Batra notes that since 1950 the world's population has more than doubled while global economic activity has quadrupled, and that between 1950 and 1990, trade grew at 1.5 times the rate of economic growth.[11] Trade is therefore a bigger polluter than industrialisation. For example, air transportation has made international trade in perishable foods possible, but at the cost of dumping millions of tons of jet fuel wastes into the atmosphere. Air-borne trade emitted into the skies 2.1 million tons of nitrogen oxides in 1990 alone. It makes no environmental sense for one firm to manufacture steel nuts in Asia, while another makes bolts in Europe and a third firm puts them together in the United States. Batra, an economist who had previously supported free trade, says that the environmental costs of global commerce have been totally ignored by orthodox economists.

Economic globalisation will accelerate the destruction of the biosphere.

The rapid increase in international trade has grave implications for the spread of pests and diseases. The Australian Academy of Science has highlighted the recent invasion of Australia by 19 alien insect species, many weeds and seven livestock diseases. We have failed to learn the lesson of the rabbit — an introduced species that has wreaked havoc on the natural environment. It is likely that at this moment a new species is gaining access to Australia which in 25 or

50 years will cost billions of dollars to control or eradicate. Why is this happening? The answer is that world trade regulations pressure Australia (and other countries) to lower quarantine standards. Within Australia, the quarantine service is financed by government, and it therefore suffers from under-financing and cutbacks under economic rationalism, which favours small government and the freedom of trade.

R.D. North in *Life on a Modern Planet: A Manifesto for Progress*, a work criticising the environmentalism of books such as our own, observes:

> *The 1980s and the 1990s, so far, have developed a consensus that capitalism must be left to operate with as little, and as carefully targeted, political interference as possible . . . The political support for free trade and market economics is now so great that it has margin-alised all left-wing thinking, including Green-tinged left-wing thought.* [12]

However, not all agree with North, even from outside of environmental thought. Robert Harvey, in *The Return of the Strong: The Drift to Global Disorder*, argues that 'the globalisation of capitalism carries with it the seeds of worldwide political instability'. He goes on to say:

> *Economically, global capitalism may be at a stage not far different from national capitalism at the end of the last century: its authoritarian nature, disregard for national and personal sensitivities, enormous power and incompetence could give rise to a perilous reaction. The parallel is with the complacency of the Edwardian era, which foreshadowed four decades of war, revolu-tion and economic turbulence. The major powers at*

the end of this century, as Cold War memories fade,
exude the same self-congratulation as at the end of the
last. Unless action is taken we will gaze upon the same
horizon of global horrors as our great-grandfathers,
this time through a nuclear haze. The world is prob-
ably a more dangerous place than it has been in nearly
half a century.[13]

In this chapter we have supported this vision of the modern world, against the comfortable optimism of the pro-growth, 'business as usual school' of economic rationalism. It is still unfashionable to criticise global capitalism. It has wholehearted support from governments, industry, commerce and the media. However, fashions change and we observe a growing number of critics of the so-called new world order of open borders and open markets. Trade union movements are uncomfortable with the abolition of tariffs and the often destructive effects of rapid changes in job markets. The churches are increasingly concerned about the social consequences of global markets and the subservience of human values to the needs of financial markets. In community movements that support the environment and equity in social and health matters, the alarm about economic rationalism is almost universal. Unfortunately the mind-set of our universities has become one of subservience to economic needs and only a minority have expressed concern. One Australian vice-chancellor, writing of the irrationalities of the economy, has argued – without much hope – for a correction in the balance between public and private enterprise.[14]

George Soros has made billions of dollars in the world's currency markets. He knows the capitalist system well. In a recent article he argues that an open society is threatened by excessive individualism and that too much competition and too little cooperation may lead to intolerable inequity and to instability.[15] Present day capitalism is not tempered by the common interest. Furthermore, he questions the fundamental economic belief that full and competitive markets bring supply and demand into equilibrium to ensure the best allocation of resources. George Soros may be but one lone swimmer against the tide of capitalism, but his thoughts may indicate that this tide might be turning. He concludes:

> *Our global society lacks the institutions and mechanisms necessary for its preservation, but there is no political will to bring them into existence. I blame the prevailing attitude, which holds that the unhampered pursuit of self-interest will bring about an eventual international equilibrium. I believe this confidence is misplaced.*[16]

We regard the conclusions drawn in this chapter as the key to our argument. The origins of every environmental health problem discussed in this book can be seen to originate in the economic thinking and the value systems of our society.

1. Michael White, *The Guardian*, London, 25 August 1996.

2. Paul Ormerod, *The Death of Economics* (Faber & Faber, London, 1994).

3. P. Wilson, 'Ralph Nader Targets BHP Over Ok Tedi', *The Australian* 7 September 1995.

4. *The Australian* 30 August 1995.

5. P.J. Lavers, 'BHP Unfairly Blamed Over PNG', *The Australian* 13 September 1995.

6. *The Australian* 7 September 1995.

7. Peter Fries, 'Mining Industry Must Lift Environmental Game', *The Australian* 5–6 August 1995.

8. J.C. Wiseman, 'NAFTAmath', *Arena* magazine, Apr–May 1995.

9. H.E. Daly, 'The Perils of Free Trade', *Scientific American* November 1993, pp. 24–29.

10. Wiseman, op. cit.

11. Ravi Batra, *The Myth of Free Trade* (Charles Scribner, New York, 1993).

12. R.D. North, *Life on a Modern Planet: A Manifesto for Progress* (Manchester University Press, Manchester, 1995).

13. Robert Harvey, *The Return of the Strong: The Drift to Global Disorder* (Macmillan, London, 1995).

14. Don Aitken, 'Public and Private', *Quadrant* Collingwood, Victoria, Jan–Feb 1997.

15. George Soros, *The Atlantic Monthly* February 1997.

16. Soros, ibid.

CHAPTER 7

CULTURE, NATURE AND HEALTH

As explained in Chapter 1, our common cultural values, sentiments and beliefs influence our ideas as to what is health and what is disease. We saw that Western medicine has difficulty coming to terms with persistent pain in the belly when tests and investigations fail to detect a cause. We saw that our society places a high monetary value on disability in the famous compared to the average citizen. We explained that indigenous people regarded themselves as diseased if they lost their land. In Western society, too, cultural traditions shape our attitudes to disease and our attitude to environmental issues in turn affects our health.

A medical whodunit

Gillian Towers is a 29-year-old nurse with important achievements in her profession. She decided to have a three-month working holiday before taking up a senior position in a private hospital in Western Australia. In South Africa she worked part-time in a children's hospital so that she could see the sights of beautiful Cape Town. She had a typical bout of influenza, with headaches, fever, a cough and tiredness. After a week she felt much better and went back to work. But a week later, the 'flu' recurred. She did not work again in South Africa, her holiday was in ruins, and she flew back to Australia. The 'influenza' continued. It caused her to be dreadfully tired; most of the day was spent in bed because if she was active for even an hour her limbs ached, she had profuse sweating and she felt light-headed. She could not even think about work, let alone commence her new job. Within two weeks of returning home she had seen her first specialist, an expert in infectious diseases. Tests for possible infections (such as malaria and glandular fever) and many viruses were all negative. Her blood counts were normal and the only abnormality that could support her viewpoint that she must have an infection, was her temperature, which on occasions was just above normal. She was given antibiotics, but they had no effect.

Over a five-month period Gillian' s illness became worse. Her debilitating tiredness blighted her life. She had no energy to shower, eat or think; she languished between bed and an armchair. Some days she had pains in her muscles, some days she did not, some days she was nauseous, some days she had diarrhoea. Then she began to have a feeling of 'pins and needles' in the side of her face and her arms; sometimes this feeling changed to burning. A neurologist examined her but could find nothing abnormal. Her life seemed to be in tatters, she became irritable, emotionally upset and depressed. Her friends, who visited to help her with household chores, began to shun her. She saw a psychiatrist and received antidepressant drugs; they had no effect.

As the months passed by, Gillian became stressed and distressed for another reason. She detected a subtle change in the attitude of her doctors. The initial concern and attentive attitude became questioning. 'Are you sure you can't stay out of bed for longer?' or, 'You must be feeling better today.' Her friends and family were starting to question her as well. She realised that people suspected her of malingering and her distress and isolation became deeper. All the more so because she had a strong work ethic and had never taken sick leave before.

After six months her condition was diagnosed by an immunologist. She had chronic fatigue syndrome. But this made no difference to her life, symptoms, or the attitude of her friends and doctors. She seemed to be an embarrassment to all. After two years, she is still living a twilight existence. The disease is unpredictable and can continue for years.

Chronic fatigue syndrome is a terrible disorder for any person to suffer. It never causes death but it disables by causing profound tiredness, thus preventing normal living. There are a whole range of distressing complaints which persist month after month after month. Patients are passed from doctor to doctor in search of a diagnosis. Many are assessed by welfare departments and threatened with having their benefits cut.

We are coming to recognise that chronic fatigue syndrome may be due to the release of substances called cytokines. These chemical substances are released from cells when the cell is infected with a virus. When they are released into the blood stream they cause the symptoms of influenza, particularly the feelings of weakness, tiredness and fever. The symptoms of 'flu' persist for as long as the release of cytokines continues.

However, chronic fatigue syndrome is a disease without a test to prove its existence. This is the problem. It cannot (yet) be diagnosed by scientific medicine, or by the physicians whose training is based on the biology of disease. It does not yet fit with our concepts of 'disease'. Doctors and the public have difficulty in accepting such a diagnosis. If the patient had such profound debility because of AIDS or any other recognisable chronic disease our attitude would be different – supportive and often positive, for the patient would fit into our scientific and *cultural* concepts of disease.

This chapter will explain how our Western cultural values affect our perceptions of health, the environment and nature. Our belief in scientific medicine stems from our Judeo-Christian heritage and the Enlightenment tradition, and Gillian's suffering shows the limitations of a narrow scientific view. Our inability to accept the concept of ecological health has the same root causes as our inability to accept the existence of chronic fatigue syndrome. We want things to be proved 'scientifically', and so we have difficulty accepting what lies beyond the limits of conventional science. So we need to be critical towards our common view that we should use our reason to analyse nature scientifically. This is part of a complex of beliefs that we can use this knowledge to control the natural world and to intervene, with technology, to improve human well-being. These ideas, which have evolved over the past three centuries, have been called the Enlightenment by philosophers.

The Enlightenment project and its decay

The Enlightenment refers to a family of ideas within our culture which we take for granted and unconsciously live by. Peter Gay, a major historian of the Enlightenment, has recognised that the common basis of various Enlightenment ideas was hostility to the Dark Ages and religious superstition, a belief in the critical use of reason to change the world and the liberal reform program. According to Gay its great thinkers were Voltaire (1694–1778); Jean-Jacques Rousseau (1712–78) and Immanuel Kant (1724–1804). He argued that the American Revolution of the 1770s, with its commitment to life, liberty and the pursuit of happiness, could be seen as the practical fulfilment of the ideals of the Enlightenment. The Enlightenment stood for the principle that we should be guided by reason rather than faith, superstition, revelation or arbitrary authority.[1]

The seventeenth-century French philosopher Rene Descartes expressed the hope that the common good would be attained by understanding nature, by making ourselves lords and masters of nature and by using scientific reason as our instrument to master nature. He says that an understanding of nature would:

Facilitate our enjoyment of the fruits of the earth and all the goods we find there, but also, and most importantly, for the maintenance of health, which is undoubtedly the chief good and the foundation of all other goods in this life. For even the mind depends so much on the temperament and disposition of the bodily organs that if it is possible to find some means of making men in general wiser and more skilful than they have been up till now, I believe we must look for it in medicine . . .[2]

These claims for medicine's potential were seized upon by doctors in the eighteenth century who argued that, in a rational society, disease would be conquered by scientific medicine. Scientific medicine would contribute to the progress of civilisation by making sick populations healthy.[3] This medical Enlightenment found an early expression in the founding and flourishing of the eminent Edinburgh University Medical School, and the growing use of London hospitals as sites for lecturing and clinical instruction in the mid-eighteenth century. Edinburgh offered one of the most systematic medical educations in Europe, and was influential in adopting a broadly mechanical approach to the body, with disease being seen as a structural and functional breakdown. One of the authors trained in Edinburgh in this tradition. Medical enterprise became a pillar of modern society and medical practice soon became a service supplied by individual practitioners in the open market.[4]

Scientific medicine, that is the happy marriage of medicine and laboratory science, with its mechanistic account of the body, bacteriology and the germ theory of disease, has made real progress against illness, and has made life safer and healthier. We have discussed examples in Chapter 1. However, in contrast with this justified Enlightenment confidence, there was an increasing awareness that the 'enlightened' industrial way of life had the potential to undermine and destroy human health. Since the early nineteenth century, it has been recognised that the 'civilising' process of capitalism has lead to industrial pollution, which has become a creator of disease. This resulted in some doctors becoming diagnosticians of social pathology and crusaders for social improvement through public health.[5]

We still work and think within the Enlightenment ethos. Our society's view of our intellectual history is that once science had won the struggle against religion, myth, and ignorance, the Enlightenment idea of progress took centre stage. When coupled with the expansion of material progress through capitalism, the result would be more liberty, humanised environments and perfected human beings. During the nineteenth and twentieth centuries we have witnessed a family quarrel between socialism and capitalism, a quarrel about the best way to implement progress by using scientific reason to design a more rational society. Liberal reformers like John Curtin in Australia, Attlee and Gaitskell in Britain and Roosevelt in America, held that after the ravages of economic depression and war the application of the social sciences could be used to reconstruct a more caring, capitalist society. This 'New Deal' would stabilise the capitalist market through economic management; it would reform outmoded social institutions and authority structures, and it would allow the working class to reap the benefits of capitalism's scientific control of nature through the welfare state.

But there are increasing reservations about the Enlightenment dream. The progress of civilisation through the management and exploitation of nature has not led to the promised land. The use of knowledge, science and power to bring about the best of all possible worlds has resulted in nature being destroyed, in both capitalist and socialist societies. Critical voices like Ralph Nader, the US lawyer, advocate for consumer groups, environmental activist and presidential candidate, hold that the creation of wealth is increasingly at odds with health. The progress of civilisation is the cause of our increasing illness to the extent that we are suffering from environmental pollution and environmental degradation. Science and technology, once seen as universally good, are now regarded with suspicion by some leading thinkers. Scientific medicine itself is recognised as a cause of injury to some people.

The leaders of nations and managers of industries retain their faith in the broad Enlightenment ethos, even though unlimited growth is ecologically unsustainable and carries high health risks. They have not wavered in their faith in the technological fix, in the mechanisms of the market to solve global ecological problems and in the capacity of scientific medicine to provide the required magic bullet. These leaders believe that human autonomy, freedom and well-being require economic growth, not the romantic harmony with nature advocated by the alternative lifestyle enthusiasts. The cost of this economic growth is the acceptance of the continued exploitation of nature.

If we accept these views at face value then we must conclude that there is something inherently wrong with Enlightenment thinking itself. It implies unlimited growth through the human domination of nature, yet an increasing number of scientists – who are themselves part of the Enlightenment – reject this notion because unlimited economic growth cannot be ecologically sustained. Nevertheless, the Enlightenment ideals of freedom and equality are worthy ones, and medical science has given humans the option of a better and longer life through the conquest of diseases. So we currently stand at the crossroads.

A critical faculty is a good strategy for resolving this predicament. On the one hand we must criticise the way society and nature are seen by economists solely in terms of market behaviour (see Chapter 6). This economic viewpoint sees everything – including human beings – as useful resources to increase economic growth through private capital.[6] On the other hand we desire to retain a commitment to the Enlightenment ideals of freedom and equality for all, and to use science to help us live fulfilling lives. An option for resolving this predicament would be to provide a broader and deeper understanding of our Enlightenment values. This is one that is more compatible with the aim of a full human life in an ecological sustainable society. Is this a viable option? In the next section we show that it is by examining one of the most valuable Enlightenment beliefs, which holds that progressively improving our lives is dependent on learning – on education.

Educating for eco-citizenship

Let us use a modern example to examine the role of education in creating a more rational society.[7] Consider the three ways of handling a common medical problem in our society – high blood pressure. A family doctor working in private practice or in the public system may prescribe a pill produced by the pharmaceutical industry and send the patient away; a medical expert may improve the population's health through enlightened preventative public health measures on behalf of passive and grateful clients; individuals may take charge and responsibility for their health with some advice from the doctor about lifestyle causes of blood pressure. Today the first option is overwhelmingly adopted for there are vested interests at play. But all three options have their place. It is possible that much hypertension could be avoided by community education on lifestyle, preventative public health measures would strengthen this option, whilst a minority of cases would require drugs. Thus in a truly enlightened democracy, citizens would be educated to think for themselves, define their own needs, seek to make institutions more responsive to those needs, and be able to act for themselves as effectively as possible to achieve their health and happiness.

Some progress has been made towards this goal. Since 1945

some Western countries have managed class struggle by establishing a national, hierarchical education system: kindergarten, primary and secondary school, regional university, and national, research-oriented university. Education, it was hoped, would give equality of opportunity to the working class, so they could lead healthier, better and more civilized lives.

But Western countries are far from being nations of enlightened citizens. Our systems fail to deliver in many crucial areas, including critical reasoning, education for living and caring for the environment (see Chapter 9). Instead, education under industrial capitalism has served a double function. The lower levels of the education system are often seen as vocational training for working-class people to obtain a job in the factories and service industries, and for the middle class to obtain jobs in administration or the professions. These people are trained to accept their station and duties and make the best of matching their abilities to the industrial machine. The upper levels of the educational system, with their division between the humanities and the sciences, are concerned with fostering supposedly critical thinking for an intellectual elite, who socially engineer a more 'rational' society.

Within the centralised capitalist state, education increasingly performs the function of upgrading the skills of workers. The humanist component now functions simply to integrate citizens into the virtues of our culture so that they will accept the ways things are. Education for citizenship has been forgotten. Medical education also fails to realise this ideal of the Enlightenment. The medical profession can be seen to be divided into a minority of idealists who serve the public interest, and a majority of entrenched, hierarchical experts who unknowingly undermine the Enlightenment by operating within a set of restrictive professional trade practices imbued with commercialism. The medical reformers have tended to become technocratic public health bureaucrats who paternally guide us to a better future. But they have little commitment to radical social change or ecological enlightenment.

This means that we need to be critical of the enlightenment tradition if we are to realise an enlightened democracy, where citizens can use their critical reason to acquire those virtues that would enable them to have a say in their own health, and in running their own nation.[8] We need to recover the Enlightenment notion of an educated citizenry and transform it, so that citizens are able to cultivate a worthy, ecological life, as part of a flourishing, ecologically sustainable society.[8] This is the looming conflict between Greens and the market over the options for living well. It is a battle in which the environment is at stake, and we can facilitate its resolution by developing a broader and deeper understanding of the Enlightenment tradition.

The liberal block

Sadly, our liberal political philosophy is failing to take such a political reform seriously. Those who espouse liberal values believe that the Greens are imposing the Green way of life on people. Liberals argue that as the environment is a disputed subject, the state should be neutral and not influence children into adopting some controversial way of life.[9] Contemporary liberals block the Green push for a more ecological way of living by giving preference to individual liberty over Green virtue.

But this liberty in fact confines us to a conveyor belt education for the industrial system. Choices are guided to provide good obedient workers, autonomous problem-solving technocrats and strong entrepreneurs. The education performed by the culture industry forms our identity and subjectivity, so that we desire more television sets, washers, cars, and supermarket foods, and are unwilling to own less. It educates us to dream about a better life which can only be satisfied by the market because there is no reasonable life outside the wealth-oriented global market. The culture industry does not educate us to become eco-citizens with those civic virtues that ensure that we care for nature, and our fellow humans. Happiness lies in the potential of the lottery ticket, and a lack of purpose is replaced by drugs and pop idolatry. The flawed economic system is creating an underclass of unemployed, who are then isolated and penalised for not participating in the capitalist system. Unemployment in the capitalist system means an inability to participate in any role; it should not be so in an enlightened society.

Why should freedom of choice in the market place be the main purpose of an education system? Why not build a rational self-sufficient ecologically sustainable life that would be humanely worth living? A common liberal argument used to discredit environmentalism is that an ecological crisis must be solved by continuing growth. The alternative would impinge upon the private freedom of individual choice. In other words, we could not proceed to a Green solution to the crisis for it would imperil the freedom of those who did not subscribe to it. The proponents of this view fail to understand that to go after unlimited growth, as at present, also imperils the lives of those who do not subscribe to it. If we want to retain the Enlightenment's dream of a democratic citizenry then an ecological, democratic education would predispose us towards some ways of life and away from others. As citizens it would enable us to critically judge which are the better ways of living.

We stress that we are writing about political liberalism with a small 'l'. In many countries there are Liberal political parties. In these democratic countries, all parties profess liberalism whether they are called Labour, Conservative, Liberal, Socialist, Republican or Democrat. Most of these political liberals block any substantial movement to a Green way of life which could lead to an improvement in health in its widest sense. How has this come about? A liberal philosophy consists of a family of ideas which define the relationship between the market, the individual and the state in the terms of rights of the individual against the state; liberal practice means a free commercial market, domestic and international peace, economic development, widespread intellectual enlightenment, toleration of a wide variety of religious and moral beliefs and a secular state, together with representative political institutions. This provides the basis for individuals to make lives for themselves within a framework of moral habits, a liberal way of life, dialogue and debate, and liberal virtues of critical inquiry and autonomy.

The philosophy of the New Right is part of the liberal tradition, as it argues for the priority of individual rights over common good. It turns to American liberalism for support.[1] It holds that individual rights cannot be sacrificed for the sake of general welfare because we need to respect the existence of each person's concept of the good life. Each individual is ascribed a set of rights which function as a 'no trespassing' sign, a fence erected around the individual, guaranteeing the individual's protected choices. These rights prohibit decisions by government that might impinge on the private sphere of the citizen, and the right of people to engage in their own projects and actions as they will, within certain limits.

The subconscious acceptance of these ideals explains many conflicts in our society. Immunisation against measles will prevent a serious illness in nearly all children. However, the individual parent has the right to refuse vaccination for their child; the reason for refusal is often supported by the fact that a tiny proportion of children will have severe responses to the vaccine. If enough parents refuse, then the disease will remain active in the community and some children, particularly those too young to be vaccinated, will die. Thus the rights of the individual can conflict with those of society as a whole. The situation is often different in societies not imbued with liberal thinking. Compulsory limitation of family size in Western countries would be regarded as a totally unacceptable infringement of individual rights. In China, the needs of society as a whole override the individual's rights (see Chapter 5). This 'Western' attitude towards individual rights explains our failure to accept collective measures to preserve biodiversity and prevent global warming and soil degradation. It is the individual's right to cut down and sell his forest, to develop another coal mine or to produce two crops a year to alleviate a financial loss on a falling market. The needs of society as a whole, now or in the future, gain limited acceptance. There is a different cultural attitude to individual rights in Asia, and the Prime Minister of Malaysia has suggested revision of the international definition of such rights.

Historically, liberalism has, faced with working-class demands for a share of power, argued for rational concessions such as representative democracy, the provision of welfare and national identity. Such concessions would prevent the working classes' revolutionary fervour from interfering with the process of capital accumulation, for such a revolution would undermine the structures of the economic system. This policy has been enormously successful. The struggle for change has largely taken place in accordance with the liberal agenda, and within the rules established by liberal culture, which have allowed limited democracy. So even though liberalism has defended capitalism by keeping conservatism and socialism in check, it has acted rationally. Yet there are limits. Liberalism's continuing strength is its strategy of rational reform, with its package of limited compromise and seductive optimism about the future within a global capitalist market, in the face of the stark and growing inequality between nations under the world economic system. Liberalism promotes this optimism through its mass media, which play down impending crises. The liberal agenda of the self-determination of nations, the economic development of the underdeveloped countries and the expansion of freedom through United Nations democracy is an important agenda. The liberal-engineered solution to the ecological crisis is the enlargement of the free-trade global economy. The solution ensures that the powerful capitalist states regulate the power relationships within the world economy through their supranational institutions. Liberalism's agenda of rational reform then provides a potential compromise between economic growth and the management of the ecological crisis. Its concept of the quality of life holds that some kinds and degrees of pollution are morally unacceptable – oil spills, nuclear pollution, wanton contamination of groundwater – and are to be prohibited and criminally punished; that we respect bio-diversity by saving those species that are ecologically and politically viable; that we have integrated land management that mixes land use and limits the use of natural or wilderness areas. This reform agenda has limits as the reader will note that punishment for environmental transgressions is limited mainly to nuclear and chemical pollutions which may *directly* affect our immediate health. By contrast one can vandalise our forests and receive the plaudits of economic society.

However, the promise of liberal reform is undermined by the widespread disillusionment with the promise of development and by the loss of faith in rational reforms. Liberalism is losing its historical foundation as the defence and legitimisation of capitalism. Its universal enlightenment claim, of speaking on behalf of all, is undermined because it is seen to speak mainly on behalf of the affluent parts of the capitalist world economy. Moreover, liberalism is seen as a European political strategy rather than the universal ethical and political theory it claims to be.

A fundamental environmental problem of liberalism is how it deals with the 'commons'. The term 'commons' is given to grazing land around a village, available free to all citizens. In *Tragedy of the Commons*, Garrett Hardin argued that if an individual grazed more and more sheep on the commons his economic gain for each sheep would be greater than the cost of over-grazing (environmental degradation) because this latter cost was shared equally by all.[11] Those taking unfair advantage of the system gained more than those supporting the system. If the air and sea are the global commons shared by all then there is no incentive for individuals or nations to control their own exploitative use of these commons. The cost of acid rain and chlorofluorocarbons in the air, oil spills and plastics in the ocean and the depletion of fish, whales and seals is shared by all who fish, breathe and live. The liberal solution is mutual coercion, with nations voluntarily consenting to rational regulation of use of their resources. Readers can decide for themselves whether this solution works for the world's fish stock, fresh water, timber or any other resource, given the recent forest fires in Indonesia.

Liberalism and the Green movement

One area of conflict of ideas today is between liberalism and the Green philosophy. The reader can see this irrationality expressed in the daily newspaper when they find the leading article attacking some part of the Green agenda, the columnist haranguing Green protests about environmental damage, or news about industrial antagonism to Green agendas. The New Right case – that these agendas are a major threat to the capitalist economic system in that they are preventing progress and costing jobs – is pronounced through press and television, which are owned by the human products of liberalism and paid for by every consumer. The Green response is made in a multitude of magazines and leaflets paid for out of necessity by each individual in the Green movement. Only Greenpeace's professionalised radicals have broken through the media barrier; a combination of spectacular action and civil disobedience on issues that the public care about (such as whales, oil pollution and nuclear explosion) forces the media to carry the story.

At this point, we need to define 'Green' to show how irrational the liberal block is. We are not referring to the 'Green' political parties that have arisen in many countries. Rather, we are referring to a Green way of life and philosophy upon which the 'Green' political parties draw to varying degrees. There is no precise definition of the Green philosophy any more than there is of liberalism. But the ideals of Greenness are eminently rational ones, as can be seen in the apt summary by Jonathon Porrit. He says there is:

> *A reverence for the Earth and for all its creatures; a willingness to share the world's wealth among all its peoples; prosperity to be acheived through sustainable alternatives to the rat race of economic growth; lasting security to be acheived through non-nuclear defence strategies and considerably reduced arms spending; a rejection of materialism and the destructive values of industrialism; a recognition of the rights of future generations in our use of all resources; an emphasis on socially useful, personally rewarding work, enhanced by human-scale technology; protection of the environment as a precondition of a healthy society; an emphasis on personal growth and spiritual development; respect for the gentler side of human nature; open, participatory democracy at every level of society; recognition of the crucial importance of significant reductions in population levels; harmony between people of every race, colour and creed; a non-nuclear, low energy strategy, based on conservation, greater efficiency and renewable sources; an emphasis on self-reliance and decentralised communities.* [12]

New Right liberal dismissal of these Green ideas is irrational when it explains to the public that a Green way of living is a recycling of the nostalgic relics of yesteryear, and adopts a rural alternative lifestyle epitomised by bearded men and home-birth of babies. It also believes that a common good which is Green must have totalitarian implications. (Green totalitarianism is discussed further in Chapter 11). In other words, we would all be forced to live in this undesirable way. This is irrational because the Green ideas of ecological public mindedness, civic activity, political participation, human fulfilment within a conserver society are rational ideas that remain within the Enlightenment tradition. Yet a Green life is seen by the New Right as a threat, and has produced a frenzy of literature from so-called independent institutions (paid for by industry) to counteract the ideals of environmentalism and Greenness.

Green activities, because they are based on new ideas, often threaten those with a conventional economic view of the world. We will quote two recent examples of irrationality from Australia. Tasmania is a stunningly beautiful island which prides itself on a clean, Green image, yet each year huge areas of pristine forest of worldwide significance are clear-felled and exported as woodchip. One outstanding wilderness area in the north-west (the Tarkine) recently had a road bulldozed through it to provide access for development. Young environmentalists, calling themselves Tarkine Tigers, protested peacefully and their voices were heard nationally and internationally. They erected tents on the lawns outside Parliament House in Hobart as part of the protest. This led to the conservative (called 'Liberal' in Australia) state government of Tasmania making a law to prohibit demonstration by camping outside Parliament House. The penalty: 5000 dollars or gaol. The Minister of Health made a speech in Parliament calling the young people ' animals, ferals and dirty'. What an over-reaction to young people who care about the future of the world. In an Orwellian gesture, the contractor who bulldozed the road through globally precious wilderness forest and swamp was given an environmental award.

Now it is often said that the Greens attract personal aggression because they are seen as 'way out', scruffy and occasionally unwashed. The Australian Greens party ran candidates in the 1996 national election. Australia has a system of preferential voting and the representatives of each political party stand outside the polling stations and hand out how-to-vote cards for their supporters. One of us joined this system for a day and handed out Green how-to-vote cards in a very conservative and affluent suburb of Adelaide. He was impeccably dressed – white shirt, green tie, smiling affable face – probably much more so than the representatives of the other political parties, yet unlike them he attracted verbal abuse from males of affluent appearance – 'f—k off, go and plant trees', etc. These events epitomise the enormous difficulty experienced by participants in the consumer society who are threatened by ideas, however altruistic, which might curtail their capitalistic viewpoint.

So we need to engage in a critique of the liberal Enlightenment in terms of its block towards the common good of an ecologically sustainable life and its irrationality towards a green way of life. In being critical we have remained within the enlightenment tradition, and commited to its ideals as we have endeavoured to extend and deepen this tradition so that it becomes an ecological enlightenment. It is therefore a viable option to criticize this tradition from within by showing its limitations. We are now in a position to turn our attention to evaluating the presuppositions of the way an enlightening science regards nature.

Ways of regarding nature

The Enlightenment has authorised the increasing domination and vandalism of nature because of the underlying Enlightenment assumption that the natural world is inferior to the human world. When the British strode into Australia, for instance, they defined it legally as *terra nullis* – an empty land. They were then free to take possession of it, master it and dispossess the indigenous people. Val Plumwood in *Feminism and the Mastery of Nature* argues that the Enlightenment regards nature as a passive *terra nullis* or a resource to be transformed into a civilised state for human interests.[13] Only humans, who have consciousness, rationality, creativity and freedom, can give value to nature and therefore allow it to be appropriated as private property.

The consequences of the Enlightenment's way of regarding nature can be seen within scientific medicine and science itself. This view of nature has led to our utilisation of millions upon millions of animals in experiments. Our culture tells us that we are more important than any other animal. To help us develop a cure for AIDS, other primates are injected with the infection. Primates are also killed so that their organs can be transplanted to humans. Yet science accepts that the genetic make-up of these animals is only slightly different from humans' and that they have intelligence. If we could develop methods of measuring their true intelligence we might find some apes to be more intelligent than the most mentally disabled of our human race. Would we then kill such a disabled human to transplant the organs? No, our value system tells us we always have precedence over nature. However, our value system has developed positively to pay attention to the pain and suffering caused to animals in experiments. We have ethics committees in universities and hospitals to oversee animal experiments. The public has concerns and has boycotted cosmetic products that have been developed by means of painful experiments on animals. This concern has extended to the confinement of fowls and pigs for food. Although there are moves in scientific medicine to replace animal experimentation with experiments on cells, scientific medicine is nevertheless going to be dependent for its continued success upon animals. Our view of nature as subservient will continue and this will colour our view of all nature. This will be to the detriment of our longterm health needs, which require that we enter into a partnership with nature.

The environmental philosophers Richard Sylvan and David Bennett in *The Greening of Ethics* argue that the Enlightenment gives rise to an ethical perspective which holds that it does not matter what happens to nature as long as it does not affect other humans.[14] Constraints are imposed on the way we treat nature, but because all value resides with humans, the value of nature resides in its value for humans. The most we can do is to respect nature while making it serve human interests. The problem with this position, argue Sylvan and Bennett, is that the ethical responsibility to nature is too shallow to protect nature, as it holds that the only ethical problems are those that concern human interests. Thus it would be ethically proper to dump wastes into the Murray/Darling river system or the sea until enough people became ill from eating the fish, or swimming in or drinking the water. In practice, this is what has happened!

The liberal assumption – that all individuals are separate to the environment, yet equal, naturally competitive and self-contained machines with rights – leaves no room for an ecological relationship with nature. It is a self-centred approach that results in the commons being fought over rather than shared. The failure to place humans in ecosystems with an appropriate appreciation of their role and respect for the environment, has resulted in some proponents of 'deep ecology', such as Warwick Fox, arguing that we should look elsewhere for help to resolve these urgent tasks.[15]

This environmental ethic leads to a questioning of the Enlightenment's mechanistic view of nature. Nature was seen as simply molecules in motion, energy and mass, a collection of isolated bits and pieces, devoid of any characteristics of thought or purpose and devoid of meaning or value. These ideas persist in contemporary form as the scientists George Wetherill and Charles Drake claim that it is 'literally true' that the Earth is a 'heat engine'.[16]

Is it reasonable to think of part of nature such as a rain forest, desert or a river basin as a machine such as a watch or a self-regulating system like an air conditioner? The authors believe that the answer is no. The parts of the watch do not produce one another – one wheel in the watch does not produce another wheel –

production lies outside the system with the watchmaker. Furthermore, the parts of a watch move, but still remain the same; they are unaffected by the interactions with other objects. By contrast the parts of the Murray/Darling river basin ecological system are transformed by interactions. The land has been cleared, the fauna of the basin has been devastated and heavy cropping has exhausted the soil. Salinity is closely linked to land clearing as the trees and shrubs that restrain salination no longer exist. Irrigation has resulted in salt-encrusted land whilst the river has been greatly altered by dams, weirs and irrigation channels and by pollution from nitrogenous fertilisers, urban sewerage and pesticides. Water quality has been lowered by salinity and nitrogen run-offs, affecting its suitability for town water supplies and irrigation. Farm bankruptcies and production loss are taking their toll on the local communities, which have experienced decline in population and economic activity, increased unemployment and suicide rates, and break-up of families.

If nature is 'molecules in motion', set in motion by the action of an external force, then there can be no purpose in nature. But if we accept that human beings have the interest to shape nature in the Murray/Darling river basin to support human needs, then we must accept that animals have a purpose. Indeed kangaroos, birds and humans all have the capacity for movement directed towards a goal, and are not simply set in motion by the action of an external force. If we accept a mechanistic viewpoint, the implication is that anything that goes wrong will have a technological fix; we will leave it to readers to decide for themselves whether there will be a limit to the technological fix.

The authors' alternative view of our place in the world holds that human beings are dependent on the biosphere; that there are mutually sustaining relationships between the human and non-human worlds; that human beings are responsible for the pollution and destruction of their environment and species; and that human beings need to respect and care for all ecological systems and species on the planet and for future generations. If Descartes' dream of a healthy life through science is to become a reality then our conception of ecological health needs to be placed within an ethics of care that aims to bring about a 'Green' way of living. This requires us to develop ways of linking the health of humans to that of eco-systems, and to think of ourselves as being within nature rather than standing outside it.

Beyond a mechanistic enlightenment?

We are faced with the limits of Western Enlightenment. An appropriate question to ask is: 'how can we go beyond these limits?' The Judeo-Christian tradition has yet to make a substantive contribution beyond its traditional notion of human stewardship of a managed world.[17] So we need to ask: 'can the East help the West?' Specifically, can we find resources in non-Western philosophy to help us work out a way of regarding human beings in nature and establish our concept of ecological health?

If we take this path we need to point out that there are different non-Western philosophies and that some 'Eastern' traditions do not appear to be especially useful for developing an ecological view of health. For example, classical Indian Hinduism holds that all everyday things, including the environment, are illusory and there-fore of little importance. Others, such as Chinese Taoism and Japanese Buddhism, have something to offer, in the form of an ethical framework which supplements science. Science plus ethics can help us to develop a more caring relationship with the environ-ment. Taoism suggests that we should begin not by playing the theoretician and deriving an environmental ethic by appealing to universal principles, but by cultivating an environmental ethos in our own place and time. This personal responsibility requires the cultivation of the environment as a dimension of ourselves, and we show humility and respect to nature as it helps our personal fulfilment.[18] This view of nature is reflected in the holistic approach of Chinese medicine, in both its diagnostic and medical/health care aspects.[19] Rather than seeing ill-health as disease or a breakdown in the functioning of the human machine, it looks at health in terms of our relationship with nature, rejects treating the symptoms instead of the deeply rooted endemic causes, and aims to be preventative in its approach. Unfortunately, the practice of Chinese medicine in recent times has damaged biodiversity by pursuing to the point of extinction species that might provide exotic cures or aphrodisiacs.

Japanese Buddhism places each of us into the ecological web whilst recognising our individuality. It sidesteps 'deep ecology', which collapses human beings into nature, and holds that there is no difference between the two – with the result that the human being might be considered indistinguishable from a virus or starfish. It suggests an ethics of care for those ecological relationships that constitute the self, and an emphasis on everyday life, in which we are placed. Yet all-conquering capitalism has over-ridden these views and Japanese capitalism has raped and pillaged the environment without a qualm.

Nevertheless, Japanese Buddhism provides a basis for ecological health by recognising that our relationships with nature are important for our health. The toxic chemicals that enter the environment in various ways – emissions into the air, effluents discharged into the sewer systems, affect our health because we are a part of nature. The local environment is the home in which we are our ordinary selves and so it needs caring for.

Though the non-Christian religions have something to offer those who are disenchanted with the current ethos of the Enlightenment, we should be careful here. We need to avoid thinking that it is our cultural traditions that have caused the ecological crisis, or that alternative ways of thinking will resolve this crisis. While it is true that the primary focus should be on our relationship with nature, it is not the liberal ego and its self seeking desires which has primarily caused the eco-crisis. It is capitalism, and its technological way of viewing things, which inhibits any other kind of relationship to nature. Changing our ways of regarding nature and society and the relationship between them is not enough, as Japanese Buddhism's eco-centrism has not prevented Japanese capitalism from laying waste to nature and transforming it into vast development projects. Nature has been moulded and transformed by technology within capitalism. It is in the technology of production, its power relationships, and the unfettered global market that the root cause of modern environmental devastation lies.

Conclusion

We have argued that the criticisms of the Enlightenment are rational ones and are an attempt to create a more enlightened and rational society. In arguing this case we have recovered the political dimension of citizenship in relation to the future of democracy. This is important because the predominance of the market in our thinking has replaced citizen with consumer. John Ralston Saul in *The Unconscious Civilisation* asks, 'Has industrialisation brought us prosperity?'[20] He qualifies his affirmative answer by showing that it was the social regulation of the benefits of industrialisation by the state that improved living standards and health. Saul argues that it is not the market or capitalism as concepts that we should denigrate; instead we should denigrate the worship of the market when it is put forward as the provider of the greatest good for all humans. The market needs to be re-embedded in a political community, so that its benefits are used to enable citizens to further a more ecologically sustainable and just society.

What is frightening about globalisation of the market is the erosion of possibilities for effective interventions by governments to regulate and control it to ensure that citizens live a full human life. If governments cannot intervene, then the people cannot intervene, and a form of anarchy becomes established where the anarchists are the faceless corporations, indifferent to the human suffering and environmental damage they cause. A market as a social institution does not contain all the moral values necessary to improve all life. It cannot have the self-awareness so critical for the development of a sense of self. If there is no sense of self there can be no sense of responsibility, of right and wrong, of remorse or of justice or fairness. Only people and their governments can sense these things, so it is vital that democracy be actively rejuvenated. A rejuvenated democracy challenges the hegemony of the market, and it makes sure that the costs of the global market do not outweigh its benefits, by specifying the need to include human health and ecology when decisions are made.[21]

1. Peter Gay, 'The Enlightenment as Medicine and Cure', in W.M. Barber, ed., *The age of enlightenment: Studies presented to Theodore Besterman*, (St Andrews University Publications, Edinburgh, 1967), pp. 375–386. See also Peter Gay, *The Enlightenment: An Interpretation*, Vol. 2, Ch.1 (Knopf, New York 1967–69).

2. Rene Descartes, 'Discourse on Method', in *The Philosophical Writings of Descartes*, Vol.1, translated by J. Cottingham, R. Stoothoof and D. Murdoch (Cambridge University Press, 1985), pp. 142–143.

3. Peter Gay, *The Enlightenment: An Interpretation* (op. cit.).

4. This history is explored in Roy Porter, *Doctor of Society* (Routledge, London, 1992).

5. For an account of the early awareness of the counterproductive consequences from progress, see Porter (ibid.), and Thomas Beddoes, *The Sick Trade in Late Enlightenment England* (Routledge, London, 1992).

6. This thesis of a world picture in which techno-science frames nature as a standing reserve is derived from Martin Heidegger, 'The Question Concerning Technology', in *Basic Writings*, ed. David Farrell Krell (Routledge, London, 1978), pp. 283–317.

7. The theme of the necessity of education to facilitate an enlightened society can be found in J.S. Mill, *On Liberty in Utilitarianism, On Liberty, and Considerations on Representative Democracy* (Dent, London, 1972).

8. Geoffrey B. Frasz, 'Environmental Virtue Ethics: A New Direction for Environmental Ethics', *Environmental Ethics*, Vol. 15 (1993), pp.259–274.

9. See, for example, Bruce Ackerman, *Social Justice in the Liberal State* (Yale University Press, New Haven 1980), p. 139.

10. R. Nozick, *Anarchy, State and Utopia* (Blackwell, Oxford, 1974). This has been generally interpreted as the philosophy of the New Right.

11. Garrett Hardin, 'The Tragedy of the Commons', *Science* 162 (1968), pp. 1243–48.

12. Jonathon Porritt, *Seeing Green* (Blackwell, Oxford, 1984).

13. Val Plumwood, *Feminism and the Mastery of Nature* (Routledge, London, 1993), p. 111.

14. Richard Sylvan and David Bennett, *The Greening of Ethics* (Whitehouse Press, London, 1995).

15. Warwick Fox, *Towards a Transpersonal Ecology: Developing New Foundations for Environmentalism* (Shambhala, Boston, 1990).

16. G.W. Wetherill and C.L. Drake, 'The Earth and Planetary Sciences', *Science* 209 (1980), pp. 96–104.

17. Lynn White, Jr, 'The Historical Roots of our Ecological Ethic', *Science* 155 (1967), pp. 1203–07.

18. R.T. Ames, 'Taoism and the Nature of Nature', *Environmental Ethics* Vol.8, No. 4 (1986), pp. 317–350.

19. Chung-yung Cheng, 'On the Environmental Ethics of the Tao and the1', *Environmental Ethics* Vol.8, No. 4 (1986), pp. 351–370.

20. John Ralston Saul, *The Unconscious Civilisation*, (Penguin Books Australia, Ringwood, 1997).

21. See John *Rawls, Political Liberalism* (Columbia University Press, New York, 1996), and D. Held, *Models of Democracy* (Polity Press, London, 1997)

CHAPTER 8

BIG BROTHER, BIG BUSINESS AND HEALTH

What may happen to the health of the human race if we are unable to change our ways? In 50 years' time our population will have more than doubled. How will our health and environment be then? In answering this question we dwell on economic, social, scientific and attitudinal trends, for we are tied to these just as species are ultimately tied to an ecological web of life.

Predictions

We are aware of the dangers of prediction. In a country town in northern England that one of us used to visit as a boy, an old codger would march up and down the main street with a placard saying, 'Prepare to meet thy doom.' It was not clear if he was referring to the German bombs which descended intermittently, or whether it was a religious message. However, the passers-by found this message psychologically unacceptable and the man was dismissed as a crank. There are great dangers in telling people what they do not want to know. In fact it can produce a destructive response against what you wish to protect or what you wish to educate them on. Predictions can, however, help focus the mind on what can happen — although predictions that have been wrong have been used to strengthen the case for no action on important environmental and ecological issues.

In 1991 a bet was made between the ecologist Paul Ehrlich and the 'rational' economist Julian Simon about the future prices of resources. Ehrlich said prices would rise because of scarcity, Simon said they would fall. Simon was right. But the reason for the fall in prices is complex, and as Paul Ormerod notes in *The Death of Economics*, economists are almost always wrong.[1] Even Thomas Malthus (1766–1834), one of the first modern writers to see the problem that the increase in population posed for the sustainability of resources, was more accurate in his long-range predictions than most economists.[2]

Since the 1960s or 1970s many predictions have been made that a population bomb would have exploded by such and such a date. There would be starvation and famine within so many decades and indeed public and damaging wagers have been made by eminent individuals. We do not believe that our predictions fall into these speculative categories – today the scientific data is much more substantive than previously. The scientific consensus is that global warming is occurring and that its advance may be much faster than previously thought. Thus our predictions on the spread of malaria and other diseases, based on scientific computer modelling, are likely to happen unless global warming can be arrested. That said, in this chapter we also indulge in speculations that do not have a scientific basis – some unbridled, some with tongue in cheek and some serious. All, however, have an historical and social basis. Some of our speculations relate very simply to foibles humans have demonstrated for thousands of years – we see no evidence that they will suddenly stop now. We refer to the consequences of greed, exploitation and war. We also build on what we see as a worldwide trend towards more divisive societies, brought about by economic rationalism and globalisation. The purpose of our speculations is to remind the reader of the likely endpoint if these trends continue in their present form.

Some writers and philosophers have been quite good at predicting the future. The authors have re-read with interest George Orwell's predictions about social and living conditions in *Nineteen Eighty-Four*, written in 1949.[3] Orwell recognised the obscenities of Stalinism before most of us and described them vividly; they did not all come to pass but many did.

One frightening aspect of Orwell's description was the control of people's thoughts and minds through the Thought Police and Ministry of Truth and their slogans which used 'double-think': War is Peace; Freedom is Slavery; Ignorance is Strength. Today, the use of particular words often hides the meaning. In Australia, for example, the name 'Forest Protection Society', which you might interpret as 'a society to protect natural forests', is actually the name of a society set up to protect the use of forests for the purpose of logging.

Most frightening of all in *Nineteen Eighty-Four* was the tele-screen, sited in every building, which produced a constant series of messages both to brainwash and to watch the individual. Everywhere were posters with the face of the leader and the caption, 'Big Brother is Watching You'. Society consisted of the Leaders, the Party and the proles. The function of the ordinary human had been reduced to that of a worker bee. On re-reading *Nineteen Eighty-Four* it is easy to see it as a description of the USSR and East Germany. Most of what Orwell predicted, happened – even down to children spying on the adults. But it is even more useful to re-read the book and to ask whether any of these predictions can be applied to our present free-trade, capitalist society.

Instead of total control of thought processes, we will find that our thinking is greatly manipulated. The psychology of con-sumerism is the main thread of nearly all information received by the average person. It is provided by the 'tele-screen' which fortunately does not yet watch us, but computers do keep a surveillance on many of our actions. Our tele-screen, the television, is controlled by a handful of powerful individuals and empires worldwide, which also own and control some of the peripheral media, such as the newspapers. These media barons also unduly influence our governments. People who believe that consumerism is endangering the environment and that the environment has a finite life will experience difficulty in getting their ideas promulgated by this information system, because what they are saying is not in the media owners' interests.

In contrast to the characters in *Nineteen Eighty-Four*, we are apparently free. But dear reader, ask yourself whether you are *really* free and able to change the system within which you live – or is your 'freedom' really only impotence? We might also ask ourselves, particularly those interested in the environment, whether Big Brother is really watching us. Big Brother in our society is big government and big capitalism. Those who speak the truth are often heavied by the law, backed by the unlimited funds of the powerful. Those who work in Big Brother government, in public services around the world, may speak the truth if they are prepared to lose their jobs. Our society can sustain the Ralph Naders of the world, but will it be able to evolve and reform to meet tomorrow's needs?

This is how Orwell described the Ministry of Truth:

> *There was a whole chain of separate departments dealing with proletarian literature, music, drama and entertainment generally. Here were produced rubbishy newspapers containing almost nothing except sport, crime, and astrology, sensational five-cent novelettes, films oozing with sex, and sentimental songs which were composed entirely by mechanical means on a special kind of kaleidoscope known as a versificator. There was even a whole sub-section – Pornosec, it was called in Newspeak – engaged in producing the lowest kind of pornography which was sent out in sealed*

*packets and which no Party member, other than those
who worked on it, was permitted to look at.*[4]

As for the proles:

*Heavy physical work, the care of home and children,
petty quarrels with neighbours, films, football, beer,
and, above all gambling, filled up the horizon of their
minds. To keep them in control was not difficult.*[5]

Sound familiar?

We hope that in highlighting Orwell's predictions we have
illustrated why, in considering the potential damage to our health
due to environmental damage, it is necessary to also consider our
belief systems, our education, and the powerful in our society and
how they operate.

The future of Western scientific medicine

In Chapter 1 we acknowledged the success of Western scientific
medicine for a proportion of the world's population. This book is not
about these successes, but we need to discuss them because of the
profound positive and negative influences that flow from them.
Every scientific discovery has positive and negative consequences,
the ultimate example being nuclear fission. At the present time its
negative components are possibly greater than the positive ones – the
threat of nuclear war, the problems of nuclear waste disposal, and the
huge loss of health and land from one major nuclear accident, which,
spread over hundreds of years, may make the worldwide cost of
nuclear power greater than all its gains.

We will consider three different aspects of scientific
medicine to see where they will lead us: genetic engineering,

the war against infectious diseases, and organ transplantation.

The study of human genetics holds the promise of cure for many diseases. Consider cystic fibrosis, well known to the community as a cause of childhood diarrhoea, malnutrition and lung disease. It is a disease of white populations, occurring in about one in every 2000 births. One in every 25 people carries the gene for cystic fibrosis, but the gene is recessive – that is, it does not lead to disease unless the partner also carries the gene and the child inherits both genes. The gene for cystic fibrosis has been discovered and it can be tested for in individuals. So, there is a means of giving parents the option of not having children if the risk is present. The choice can be exerted through contraception or, in the future, by detection and abortion of an affected foetus. In the future there will be 'gene' therapy; children with cystic fibrosis will receive a normal gene to replace the damaged gene, and will become normal. This sort of treatment will eventually be possible for other genetic diseases, including many forms of cancer. This is the credit side. Other diseases will not be treatable in this way because they are caused by defects in many genes, possibly arising from environmental factors such as irradiation.

The Human Genome Project is an international program of research which aims to describe the position and structure of the 100,000 human genes on their chromosomes. These genes transmit all genetic information from one generation to the next. This 'map' of genes will facilitate the treatment of many genetic disorders. But there may be a debit side to the project. It may not lead us to health and happiness any more than nuclear power has. Confidence in the beneficial use of these discoveries depends on one's view of the social maturity of the human race. This confidence is dampened by the origins and proponents of the Human Genome Organisation. The project originated with the military, presumably so that humans could be given genetic protection against chemicals used in warfare. At a conference in 1989 David Koshland, editor of the US journal

Science, was asked why the vast funds given to the Human Genome Project should not be given to the homeless. He answered: 'What then people don't realise is that the homeless are impaired ... indeed, no group will benefit more from the application of human genetics.'[6] So there we have it, social engineering, the imposition of conformity, the answer to behavioural abnormality. The poor, as well as the eccentric – perhaps even the van Goghs and Beethovens – need to be engineered to conform to the needs of capitalist scientific society.

The power of this genetic medical science will rest with the multinationals; the discoveries will be patented, and those who can pay will benefit. Those with untreatable genetic disorders, who cannot be cured or cannot afford to be, may live in the genetic ghettos, unemployed and uninsured, their genetic ratings, like their credit ratings, available to all on the worldwide computer net. The genetic ghetto will grow in line with the trend towards decreased social support in Western culture. We are proceeding to a society with a few rich and more poor, and with large sections of society discarded as were George Orwell's proles. As J.K. Galbraith has said, the USA (and other major democracies) are increasingly developing a 'culture of contentment' which is a revolution of the rich against the poor, operating under the compelling cover of democracy. The situation has become even more polarised in some countries, for example Brazil, where the majority of people accepts that street kids can be exterminated.

Lee lives with her two children in a company collective on the periphery of a large city. Its location is immaterial, for all cities worldwide are much the same in 2050. However, the collective is greatly protected from the unemployed and neglected parts of the city. It is secure and 'special'. Its inhabitants have made significant progress by including themselves in a program of auto-evolution whereby they receive genetic attributes. Lee, like most employees, works at home and her tasks are set clearly by instructions that appear on her computer screen when she clocks on each morning.

Her two children receive their education at home from their own computer screens but two days a week they are taken to 'school' for 'social interaction'. This morning they have been collected by the security monitor for transit to the school. Lee has a terrible headache, it was present yesterday, but today it engulfs her. She reports it to her employer and is asked to put her personal disk into her computer. This contains her life-long illness record, work record, drug reactions and more. A questionnaire comes up on the screen. When did the headache start? Does she have nausea? Is there a problem with vision? How severe is the pain? Is her throat sore? Each question is qualified and broken down into possible responses. She supplies the answers, which takes her 15 minutes. She receives an instruction – arrive at Glob-care at 10.15 am. Glob-care is one of six international medical companies which manage the health of the two billion persons employed by 14 international companies. The service and expertise is offered throughout all the developed countries.

Glob-care is centralised within the urban technopolis. At reception, Lee's identification and health history are checked from the disk on her wrist and she is processed quickly through a magnetic brain scan; her temperature and blood counts are taken for infection screening. She does not see a doctor, it is not necessary; the tests are normal and 30 minutes later she is on her way home with painkillers for her headache. Medical instructions are already on her computer screen at home. She is not to work for the rest of the day because of the sedating effect of analgesics – start work the next day if no further headache – and there is a series of questions and advice on the position of her computer chair, resting from her screen every five minutes, and so on.

Lee appreciates the company's effort on her behalf. She is one of only 40 per cent of persons in work in developed countries. Her 'health' charges are deducted from her salary. She is thankful that she did not have to go to the public institution, Munipax, when she was ill. Recently an unemployed relative had an abscess. She told

Lee that the reception at Munipax was like a busy airport arrivals hall. There were hundreds of shabby citizens milling around and forming queues at the appropriate booths. Those who couldn't read were put on one side; the literate followed signposted channels and arrived to see a team of workers led by a Unidoc. Each person was assessed and a decision made on whether treatment could be given. Only prescribed and costed treatments were available. There was nothing like the personal attention and instruction that Lee received from her employer.

But Lee's headaches continue to occur every few days despite the decision that there is nothing scientifically wrong with her. She wishes she had someone to talk to about it. She finds an old-fashioned community service run by the unemployed in their ghetto and for a donation receives sympathy and support and some ancient herbal treatment. This helps her symptoms; her father had died a month before and she feels remorse at not spending more time with him. Her symptoms were due to this remorse and the conflict and stress it caused.

Western scientific medicine will increasingly embrace the information and communications systems. It will be increasingly impersonal but will serve a proportion of the population well – when they have a specific illness. Information systems will continue to revolutionise the capitalist world but they will widen the economic gap with the 'developing world'. On the day in 2050 that Lee attended Glob-care, 300,000 people died from cholera in the conurbations of South-East Asia after contamination of a river system during vast floods.

Overwhelmingly, the impact of Western scientific medicine will depend upon 'rational' economics. Those who can pay will be able to partake of any medical advice. The rest will be rationed. This 'rationing' is already carried out under the guise of 'efficiency' and protocols for 'standard care'. A system in Australia called Case-mix places an 'economic' reimbursement on each patient treated, with the aim of fast throughput, quick discharge and standardised care. The system has proved efficient for cutting hospital budgets and reducing the length of stay of patients in hospitals, but the new language has been successful in warping the issues in true Orwellian style. The final seal of approval has been the establishment of professorships of Case-mix medicine, perhaps a sign of the economic corruption of some universities.

Based on present trends, we can see that we are heading towards the world in which Lee lives. Bio-medicos will engage in eugenics and sanction it in the name of enlightenment perfection; we will try to control our evolution; sperm banks will become supermarkets; gene repair shops will be found next door to The Body Shop; and surgeons will replace 'diseased' genes with 'healthy' ones. The bio-politicians and philosophers in the education industry will assure us that nothing can go wrong as safety procedures can regulate the hazards, that all talk of monsters and eugenics is ludicrous and only serves to stir up the irrational fears of a distrustful population. Nevertheless there will be a lucrative underground cloning industry, fostered by the rich, powerful and egocentric who need to ensure the survival of their particular genome for posterity.

Social problems will be solved by genetic techniques: the modernist dream to rationally plan and realise a life without problems will be nourished by the expanding genetic possibilities. As we have seen, the Human Genome Project seems committed to this already. The technocratic philosophers will continue to defend their mechanised view of the world in the name of reason, science and truth. The Greens, they will argue, deepen the darkness of mysticism and superstition with their cuddly metaphysics with its intrinsic value of nature.

The Genome Project will reinforce the belief that humans are machines and that any defects they develop can be corrected. The scientific medicine practised by the bio-medical industry will continue to define health as freedom from disease, as bio-ethical philosophers argue that whether one is ill or well is a matter of scientific fact. As the doctor alone has the expertise to diagnose illness, it is the duty of the patient to submit to the reasoned analysis of the doctor and obey the orders given to the technicians. A value judgement is being made: that the doctor is a bio-engineer, that the patient is a machine, that illness is a mechanical failure, and that medicine is akin to repairing a malfunctioning car.

Bio-ethicists employed on performance contracts by the Eugenics and Social Correction Clinic in the university owned by Just Genes International will write papers on the question: what makes life valuable? The bio-ethicists are saying that we must answer this question first, that we must know what makes life valuable in general before we know whether this or that particular life is valuable. The next logical step from that premise is that the term 'person' designates the possessor of a valuable life, so to be a person is to have the capacity to value one's life, or to have a desire to go on living. Self-consciousness is the necessary condition for personhood, therefore anything that lacks consciousness of itself as an independent centre of consciousness existing over time, cannot be a person. This universal rule determines whether individual people are to be deemed persons: if they are not, then there can be no rational grounds for moral qualms about experimenting on them or recycling their parts when they break down. In some countries prisoners are 'non-persons' whose body parts can be recycled for transplantation purposes.

We can speculate that as the ecological crisis deepens, tourist agencies will sell mystical escape holidays from an overpopulated, industrialised, polluted, anxiety-ridden and violent world. Large parts of the population outside the privileged centres of Technopolis will lead lives that are nasty, brutish, short and sick, like the proles in Orwell's *Nineteen Eighty-Four*. The poets and nature enthusiasts will win literary prizes for their post-historic primitive texts on the classic theme 'in wilderness lies the preservation of the world'. Disney Corp International will humanise wilderness by manufacturing wild lands for technologically jaded tourists, and the successors to BHP will continue to mine and lay havoc to the national parks, which have been defined by the state as a valuable resource to solve the economic crisis. The old folk will recall a time when they were in touch with the ancient rhythms of the Nullarbor, as they spin narratives about returning to the foundations of life in the landscape, where they could relax and ponder the meaning of existence. They will call for an education system that educates people about the inherent value of wild lands and life and defends the idea of wilderness.

Lee's world of 2050 will have seen our economic system continue to damage nature, plunder Eastern Europe and lay waste to Asia with millions of peasants being uprooted in China. Its continued necessary expansion will be centred on information technologies and the standard automation of production and management. The Asian nations will continue to be technologically guided by Japan in a flying geese pattern as the advanced capitalist countries increasingly take the 'Brazilian' approach – exterminating street kids and building fortified compounds for the rich. The city of the future, Los Angeles, divided into rich and poor sections, will have been followed by Madrid, Miami, Kuala Lumpar and Sydney as the cosmopolitan ruling class retires into its own private high-tech nation within the shell of the nation-state, with its private neighbourhoods, schools, police force, health care and roads. Behind the walls of their technologically artificial world they will conduct their global business with the ruling classes in Indonesia, China and Japan and refuse to pay taxes for public amenities for the milk-bar economy outside their own territory. They will retain a high-tech military to keep law and order in the polluted and destroyed world outside, as the rising resentments caused by economic decline, lack of water and increasing pollution will be expressed as hostility between the groups at the bottom, rather than rebellion against the top.

Such are the possible scenarios for the ecologically damaged world of 2050. As a dystopia it may be a caricature but it is not a joke. It represents the bitter process of the working out of a self-negating enlightenment project, as it turns into a hollow parody of itself with its high-tech barbarism which fatally wounds people and the Earth on which they dwell. It leaves philosophy in the melancholy role of rag picker, sorting through the refuse, salvaging the debris, then stitching and patching the salvaged scraps to sew our dream together. This is an act of care as well as of mourning.

Scientific medicine will also sail into uncharted waters with its transplantation programs. The use of human genes in animals will enable animal organs to be transplanted freely into humans without present-day immunological treatments to prevent rejection. Transplantation will increasingly be a treatment for the rich with death the option for the poor – this is already the norm in American medicine. However, in the next 50 years we speculate that transplantation will be either restricted because it may spread infections or if it is not restricted it will cause new plagues of infectious disease. If, as some scientists believe, the HIV virus was transferred to humans by a polio vaccine prepared in monkey kidneys, then the precedent is set for other slow viruses to be transferred from animals to humans. Slow viruses do not produce symptoms for five to 20 years, during which time the infected human may transfer the infection to many other humans. Will it be possible to adopt surveillance programs for every transplant patient for 20 years? Delay in the manifestation of symptoms is also a characteristic of prion diseases such as Mad Cow Disease. It is likely that these diseases will increase because of transplantation, and because of poor ecological practices in the food industries.

If international actions are inadequate and if present trends continue, we predict a resurgence of infectious diseases. The major causes will be global warming and the population explosion. In 50 years' time malaria, dengue fever and many other tropical infections will have spread north and south to regions that are at present temperate. Both the malarial parasite and the HIV virus give every indication that they will be elusive to treatment and to prevention by vaccines – just like influenza. The pharmaceutical industry will have lost the battle against infections resistant to antibiotics, and deaths from infections in developed countries will have increased greatly. Human nutrition will have become less varied, even in those countries which retain their affluence, because our methods of producing chicken, beef and other animal proteins will have led to epidemics of infection and a return to more acceptable methods of production. Fish will be increasingly rare because of overfishing, and an increase in trade will have transferred one country's pests to every other. This transfer of pests will make fish farming increasingly unstable.

Global warming

The threats to our health and well-being brought about by global warming have been discussed in Chapter 4. We predict – and this is a prediction, not just a speculation– that global warming will be significant within 50 years because this warming is largely dependent on carbon dioxide pollution, which has occurred already. Can we change this course of events? There are signs that the problem is being recognised. Attempts are being made to place international legal agreements on greenhouse gas emissions but the hard-fought-for reduction in emissions by Western countries will be insignificant compared to the rise in emissions brought about by Asian 'growth'. These developing countries are not going to take the problem seriously until they have 'developed'. In fact, as pointed out in Chapter 4, to 'develop' to Western levels will mean the end of our way of life as we know it. What hope is there?

The global discovery of oil is falling and this may speed the development of alternative non-polluting energy resources. Warning calamities may occur. It is human nature to act upon calamity. A recent example is provided by the acceptance of strict gun control laws in Australia. Political debate over several decades had failed to provide any reform, but when a single gunman massacred 35 people, strict laws were agreed upon within two weeks. Hopefully there will be some warning of environmental calamity. Perhaps it will be through the extremes brought about by climate change, which may bring spectacular flooding and inundation of some lowlands. Perhaps it will be economic chaos, brought about by the increasingly unstable international financial markets, that will provide the opportunity for a new beginning. Humanity may recognise that whilst it can do nothing about earthquakes and volcanoes and other natural disasters, it will have to act to prevent further disasters caused by human intervention. The authors accept the human ability to rise to the occasion, act in the face of adversity and develop solutions undreamed of today to solve some of these problems but we do not believe that this will happen unless we change or at least reform our present imperfect democratic system and consumer society, the basis of environmental damage.

The words of E.O. Wilson, a Harvard entomologist, summarise the situation:

> A study of nature, has brought into clear relief the following paradox of human existence. The drive towards perpetual expansion – or personal freedom – is basic to the human spirit but to sustain it we need the most delicate knowing stewardship of the living world that can be devised.[7]

We believe that we are light years away from such stewardship and to attain it we will need to completely reform our present educational and economic system. Whichever health and environmental problem we analyse, inevitably we return to the common denomination of our problem – economic thinking.

Globalisation and capitalism

We believe that the continuing expansion of the capitalist economy through globalisation will result in politics and culture being shaped entirely by economic activities. Everything is increasingly seen from the perspective of the market – whether something hinders or supports the market will be all-important. Accordingly, the state will become concerned mainly with the administration of economic activity as a self-regulating mechanism. We will become just consumers, administrators and producers in the market, and the expansion of individual freedom and autonomy will take place within the form of relating to one another through contracts made to mutual advantage. The political, cultural and familial aspects of our society will be incapable of defending themselves from their colonisation by the market. The consequent loss of our shared world of experience, action and speech will diminish our highest humanist values. These will be replaced with a common consumer culture of Reeboks, McDonald's and Coca-Cola, which our culturally informed humanist passions will revolt against. We will become 'schizophrenic' in the iron cage of the market as a high-tech global economy continually mechanises everyday life. Nature will continue to be remoulded by society, and then artificially 'renaturalised' and administered as arks for endangered species and flora. Wildlife parks and sanctuaries will flourish and become living museums in the future – relics of the natural wilderness of the past. Animals will become gigantic meat factories on four legs as they are infused with synthetic growth hormones. Capitalism will colonise the raw material of human bodies through genetic engineering and the bio-construction of human beings for profit and the return on investment to shareholders.

In *Visions of the Future*, Robert Heilbroner, the eminent economist, provides us with the crux of the problem:

> *At its core, capitalism is a social order that marshals and expends its energies in pursuit of capital. Economists have differed over the source of that continuously pursued reward, but there is no doubt that expansion is a life process of the system. Without it competition would continuously erode profits, and capitalists would find no incentive for investment . . . But an economy without growth would be as incompatible for capitalism as a society without serfdom with feudalism.*[8]

He then goes on to predict that capitalism will not last forever. It will fail because in rapidly 'advancing nations' unemployment will have to increase because it is part of competitiveness and job shedding. A nation that avoided job shedding would suffer inflation and would fail to be competitive in the global marketplace. Unemployment is a prescription for social disorder, revolution and war. He then argues that if we are to retain the present capitalist system, we will have to have an increased public sector with government doing the essential jobs for society, including environmental renovation, in contrast to governments' cost-cutting exercises so ideologically prevalent at present times.

Capitalism and globalisation are tightly linked. Capitalism needs globalisation for its 'growth'. Globalism will accentuate all the consequences of the free market through the spread of pests, degradation of soils, loss of capital for infrastructure and unemployment. These will be difficult to change or modify. Of the millions of words written on globalisation, there is little mention of the environment and no mention of health and well-being. If Lee was not working for a multinational company in a developed country what would be her health experience in a 'developing' country? In 50 years there will still be 'developing' countries. Indeed, if present trends continue the gap between developed and developing will be even greater than today. The Asian economic boom will have slowed, leaving a legacy of overpopulation, pollution and environmental degradation. Today one billion of the world's five billion people earn less than a dollar a day. The evidence suggests that it will be two billion in 2050 for there are no foreseeable mechanisms to correct this poverty. In 2050 Lee's lot in a developing country will be intermittent work for low wages in the factory of a multinational in the vast urban sprawls. The countryside will be ravaged by monoculture in vain attempts to increase food production. People will be constantly migrating for work within countries and sometimes between countries. Lee's health if she lived in this 'developing' country will not be materially different to today: anaemia, malnutrition, parasitic disease and the exhaustion of frequent childbirth. The agricultural trends that lead to this scenario will be similar to those those that followed the repeal of the Corn Laws in the UK in 1846. Until then agriculture was protected and farming employed a significant proportion of the population in a stable relationship with the land. Repeal of the laws exposed agricultural production to outside competition. Prices fell and labour moved, as intended, to the factories. Urban poverty exploded.

We may have to face the message that our economic system cannot ensure ecological sustainability. What then? What will happen in the final showdown between the forces of money and the forces of life? There are many possible scenarios. The human population may be the first to be culled by war – especially nuclear-war, starvation and disease. Countries rich in land and natural resources could be invaded by countries suffering from over-crowding and shortages of resources. Civilisation as we know it – capitalism, science, liberalism, democracy – will not survive such disorders, and humans, if they survive, will become just another animal species, living in tribes and scratching out a living on the margins of existence. David Price sees it as follows:

It may prove impossible for even a few survivors to
subsist on the meagre resources left in civilisation's
wake. The children of the highly technological society
into which more and more of the world's people are
being drawn will not know how to support themselves
by hunting and gathering or by simple agriculture. A
species that has come to depend on complex technolo-
gies to mediate its relationship with the environment
may not long survive their loss.[9]

Population

The population growth predicted in Chapter 5 is almost certain to eventuate as are its consequences for starvation, war and the displacement of peoples. Only future plagues can temper this increase and are likely to occur in the next 50 years. One is the AIDS epidemic already occurring and which may have a significant impact on the burgeoning Asian population. A chemical plague induced by the pollution of industry may also stem population growth. Evidence is accumulating that there is a worldwide fall in human fertility and there is speculation that this is due to pollutants from chemical industries.

By 2050, unless we change our ways, wars will have increased because of population growth, though most will have been local, involving border disputes and the rights to rivers and productive land. There is a significant chance that some of them will have involved nuclear weapons. Why we are being so pessimistic? The answer to that is that people have carried out warfare throughout history and there are few initiatives to prevent this. There are no new international structures to help with the problem and there is increasing international competitiveness, which will ensure conflict.

Humanity does not yet have the ability to examine the inherent evil within itself. Under certain circumstances it may well be that any of us could have been a camp guard at Dachau. The horrors of the Second World War and the Holocaust are only 50 years behind us, yet there has been no let-up in war with its extermination, ethnic cleansing, rape and torture. Iran, Iraq, Bosnia, Cambodia, Rwanda are not part of our everyday mind-set or culture. Yet they *are* part of all our cultures in Europe, Africa, Asia and the Americas. Increasingly, billions and billions of dollars are spent on arms which are a huge budgetary item even in poor countries; this arms race increases as the world divides into smaller and smaller units and nationalism rises. Territories may be taken over by the most powerful and populous nations which will, in effect, colonise other countries. War will come as a result of population pressure, and the need for water, land and resources. These coming wars are likely to have a severe environmental effect because of their nuclear and chemical capability to contaminate and because the countries involved have no margin of safety to play with; the environmental degradation brought about by war will immediately compound their restricted water and food supplies.

Population growth will continue because we find the issue too tough to tackle. Sir Joseph Banks, a botanist aboard the HMS *Endeavour*, records in his journal one of the first encounters between the Aborigines of Australia and the British explorers. As the *Endeavour* sailed into Botany Bay, the ship passed four Aboriginal canoes in each of which was a man spearing fish. Banks wrote that The ship passed within a quarter of a mile of them and yet they scarce lifted their eyes from their employment'. The ship later anchored opposite a small settlement. Women and children looked at the ship but did not express surprise. Robert Ralph, in an article in *New Scientist*, asks: 'Is it possible that the Aborigines were looking at the ship but, in some sense, not seeing it, unable or unwilling to take in something so far beyond their everyday experience?'[10]

This incident illustrates the capacity of humans to become so locked into the minutiae of everyday existence that they lose the capacity to perceive possible danger. Modern society is in a similar position with the population crisis. As long as the human population continues to grow, energy and resource conservation is ultimately futile. At a growth rate of 1.6 per cent per year, a 25 per cent reduction in resource use would be eliminated in just over 18 years.

Conclusions

There are no initiatives afoot which will prevent world population at least doubling in the next 50 years. We are all agreed that this is likely to happen, only its consequences remain debatable.

Without a change in political systems and individual attitudes, the global market will have irretrievably damaged much of our remaining environment. Certainly most of the wilderness areas will be gone. Global warming will progress. The level of global warming in 50 years will be determined by what we do *now* to prevent it. And we are doing very little. So in 50 years there will be climatic instability and higher global temperatures, and the spread of tropical disease northwards and southwards will have occurred.

There is now enough evidence to indicate that our ecological systems are being dangerously disrupted. In stating this we do not base our judgement on a reduction of biodiversity in rainforests. Rather we base it on the global picture of antibiotic resistant organisms, the rapid spread of species ('pests') to new environments through global trade routes, the gross ecological changes brought about by the timber trade in Asia and the Americas and the changes in disease patterns owing to over-population, poverty and global warming.

Present trends suggest that society will have become more divided and polarised. Consumerism, profit and the market economy will lead to this – indeed it is already happening. Our present democratic system facilitates election of those working for vested interests, and a jackpot mentality. Elections are not won on provision for the poor and the maintenance of an environment in which our grandchildren can live. The media will facilitate these divisions for the media has emerged as the moulder of our ideals.

Having presented our predictions of what the Earth's fate might be, we turn now to the ways in which the ongoing and potential catastrophes of the next century can be alleviated.

---

1. Paul Ormerod, *The Death of Economics* (Faber & Faber, London, 1994).

2. The orthodox view is that Malthus failed to foresee how technology and industry would increase yields and sustain expanding populations. Yet Malthus was almost correct. He predicted that food production in 1988 would not be more than 7.5 times what it was in 1800, so that the world's population would not be more than 7.5 times what it was in 1800. In 1800 the world's population was approximately 900 million and in 1987 it was five billion, an increase of only 5.5 times the 1800 statistic. See D.B. Luten, 'Population and Resources', *Population and the Environment* 12 (1991), pp. 311–329.

3. George Orwell, *Nineteen Eighty-Four* (Penguin Books, Harmonsworth, Middlesex, UK, 1949).

4. *Nineteen Eighty-Four*, Chapter 4.

5. *Nineteen Eighty-Four*, Chapter 7.

6. David Koshland, remarks made at the First Human Genome Conference, October 1989.

7. E.O. Wilson, *The Diversity of Life* (Belknar, Harvard, 1992).

8. Robert Heilbroner, *Visions of the Future* (Oxford University Press, New York 1995).

9. David Price, 'Energy and Human Evolution', *Population and Environment* 16 (1995), pp.301–319.

10. Robert Ralph's *New Scientist* article is quoted in J.W. Smith, E. Moore and G. Lyons, *Global Meltdown* (Praeger, Westport, 1998), p. 82.

CHAPTER 9

THE GREENING OF HEALTH: SOME SOLUTIONS

What has to be done to retrieve the situation? We believe that carbon taxes, pollution controls, subsidies for the development of alternate energy, research into malaria and famine relief would all be useful measures, but peripheral to the real need to change our value systems and our economic systems – if these were remoulded all necessary reform would ensue. We have chosen four fundamental areas for change: our education systems, which increasingly serve the market economy; our economic systems, which are environmentally destructive and disregard alternatives which are already available; our attitudes and motivations – we need to take a close look at ourselves; and our values and political systems, so that those in the community who understand health and environmental needs will be empowered to introduce the necessary changes.

Changes in our education systems

The authors, who have life-long service in and commitment to education, believe that our education system is largely inappropriate to the fundamental needs of humanity and will have to be altered. We believe that without this our present progression towards increasing ecological disorder cannot be halted. Let us explain the deficiencies in our education systems in the Western world.

Education should be for all aspects of living but the present system is concerned mainly with economic living. It neglects self-sufficiency, values, spirituality, human inter-relationships and, most importantly, health and the environment. Furthermore many of its concepts are based on a male dominance. A simple example of the shortcomings of education was put forward by David Orr, drawing on a speech by Elie Wiesel at the Global Forum in Moscow:

> The designers and perpetrators of Auschwitz, Dachau, Buchenwald and the Holocaust were the heirs of Kant and Goethe. In most respects the Germans were the best educated people on Earth but their education did not serve as an adequate barrier to barbarity.[1]

What was wrong with their education? In Wiesel's words:

> It emphasised theories instead of values, concepts rather than human beings, abstraction rather than consciousness, answers instead of questions, ideology and efficiency rather than conscience.[2]

Our education is still at this point. In broad terms our education is still based on human domination and management of the environment yet we can never manage the environment any more than we could (or should) 'manage' a spouse! Better to establish a partnership in both situations. Education is concerned with success and upward mobility. This has been recognised in many countries by a tax being placed on graduates. In other words, education is equated with earning-power. But if you receive this education and fail in the economic system and become unemployed, how are you equipped to live a life that contributes to humanity and society? You are not. You have not been not been taught how to live a healthy and fulfilling life or how to subsist; unemployment ensures that you are confined to subservience and unable to make a contribution. Yet there will be increasing unemployment in Western communities, where extreme competitiveness and job-shedding is likely to be ahead of the ability to create jobs. The increasing internationalisation of industry means that jobs will go where labour is cheapest and there is least regard for health and environmental issues. The restructuring by powerful corporations is now increasing profits, reducing wages and the number of jobs, and increasing the gap between rich and poor. In the USA this is being fostered by powerful lobbies. Job-shedding is supposed to be counterbalanced by new jobs in technological industries. The reader in Germany, France, Australia, Canada or the UK will have to decide whether this is so.

Let us summarise the problems with the education system in Australia. These problems are not fundamentally different to those in other Western counties. Values and direction are set in early life, perhaps earlier than we think. The seeds of competition, consumerism, economic greed and gender roles are established as early as the infant school stage, when the child is still developing basic literacy. This value system is reinforced at each stage of education. In Australia, where school education is the responsibility of the states, a minuscule proportion of the education budget is devoted to environmental education. Often we are talking about much less than one per cent of the education budget. For junior students there is some educational value in projects involving furry animals, birds and tree-planting, but unfortunately that is as far as it goes. There is little environmental education for senior school students and in the universities the meagre resources put into environmental education are usually the subject of argument about which existing departments should grab them – usually the science departments win. The examination system, which determines whether students will proceed to university, has no room for education in values and environmental issues. The high marks are needed to gain entry to the high-earning professions such as medicine, and these marks must come from the study of maths, physics and chemistry. The environment and health are difficult to encompass in present school and university curricula.

The curriculum is compartmentalised, whereas health and the environment are part of many subjects. Environmentalism cannot be taught in the manner of other subjects, by the delivery and receipt of information; it requires experience of the environment in its natural state and the on-site examination of problems caused by humans.

The ethos of our universities today serves what are seen as the economic necessities of the nation and the universities themselves increasingly resemble corporate empires with liaisons and funding from industry and business. Even within the research sphere, national funding is defined by priorities which have had political input. We are moving steadily away from many of the essential tenets of universities such as the conservation of knowledge, the advancement of knowledge by basic research, the refinement of knowledge by critical review and scholarship, and the assumption of the role of a conscience and critic of society. Professor Ian Lowe defines these essential tenets very well in *Our Universities are Turning Us into the 'Ignorant Country'* and it is apparent that even in terms of serving our present economic system (with which the present authors do not agree) our system is failing.[3] The universities have become stamping grounds for political power-play through multiple reorganisations. Administrations have become subservient to this political play. Lowe looks at how a government waste-watch committee targeted research programs such as 'Motherhood in Ancient Rome'. Rather than try to expose the dishonesty of this campaign against research projects, university administrations did everything to escape attention short of hiding under their desks. Apparently there was no serious attempt to defend the role of research in the modern university or the importance of fundamental research in such areas as the changing patterns of parenting.

We believe that there is nothing more important to the survival of our society than a critical evaluation of our social interaction – an evaluation which may well include examining motherhood in ancient Rome – with a view to improving our social needs such as avoidance of crime, and the establishment of well-being. There are important issues confronting us, such as the social control of our technology – which is in fact out of control. Lowe continues to point out how specialisation in universities stops the best brains dealing with important issues. He chooses global warming as an example. To understand the process and its complications requires the interaction of many disciplines – physics, chemistry, biology, mathematics, computing, engineering, economics and social psychology. It is easy to see how the structure of a university could prevent a cohesive approach to the problem.

When we proceed to national and international education policy, we find that there is little understanding of the issues. Politicians are committed to the existing system and the most depressing discussions that the authors have ever had have been with some ministers of education. Their views on learning in general tend to produce total despair amongst those who listen to them. There is only slight international recognition amongst educators and leaders that there are problems that we cannot solve with our present education systems. Agenda Twenty-One, the 1992 conference on environment and development, indicated that education is critical for promoting sustainable development and improving the capacity of people to deal with environment and development issues. It was felt that both formal and non-formal education are indispensable to changing people's attitudes so that they have the capacity to promote sustainable development. A series of proposals was put to governments for implementation but there has been minimal action worldwide for the education systems are monolithic and imbued with economic rationalism. They will not be changed easily.

What do the authors suggest as change for the better? Each member of humanity has a brain, of unbelievable complexity, which can analyse our deficiencies and devise educational reform. There is worldwide a nucleus of thinkers who are concerned about today's education systems and have the ability to change them. On paper, there are excellent programs for environmentalism and environmental health but they are not implemented because of bureaucratic or government resistance. In our view there must be two points of attack: education in the five- to eight-year-olds, and the university system. (Reform of the university system will eventually determine what happens between the ages of eight and 18 because the education of this age group is designed for university entry.) Junior education programs children to be good consumers and competitors and this is reinforced by the impact of our information systems, which are largely based on television. A fundamental re-think is required if our next generation of managers, consumers and political leaders is to be any different from those of today. Australia is a wealthy, peaceful, successful, multi-cultural country. It could offer leadership in the world in so many ways.

The child's early relationship to its environment is instinctive; the child feels connected to other animals and revolts against killing and eating them. Destruction of an environment is devastating. We all have these experiences though they become suppressed and forgotten as our cultural domination over nature is indoctrinated. If we could make a single educational reform to arrest the destructive impact of humans on the world, it would be to preserve the links to nature instinctively felt by the young child. At the minimum this has to be an option for the thinking child, an alternative to indoctrination into our consumer value system.

How could we reform the universities? The authors are pessimistic because the ethos of our universities is set in concrete and they are full of people dedicated to their own self-aggrandisement. We have seen private universities established — again for the personal aggrandisement of the donor and for the worship of the present economic system. It would require only one per cent of those individuals who have become wealthy and wish to donate, to create free-thinking universities which could offer a reform model. Such enlightenment on the economic road to Damascus has occurred before and has led to benefaction.

We are not alone in our emphasis on education. We have looked at the words of Sir David Smith, former Principal of the University of Edinburgh and now Master of Wolfson College, Oxford. He is a scientist by training, who in his text *How Will It All End?* spells out our future course in education – which the reader will realise was a focus of this book. On reading Sir David Smith's book one utters a cry from the heart as to why we have not produced more vice-chancellors like this one. He seems to have escaped the morass of muddy thinking and subservience to business that has infected most of the universities in the West.

Like us, Sir David Smith is pessimistic about many of the issues discussed in this book and in particular about humankind's personal and political will to stop global warming. He agrees that:

> *The environmental bandwagon has little impact on politicians' work save perhaps on their rhetoric even at an international level. There is a preoccupation with short-term issues such as the interminable recessions and unemployment brought about by our irrational economic system and these issues swamp the essentially longer-term nature of serious environmental issues.* [4]

Given that he is an educator, Sir David Smith's conclusion is not surprising:

> *Essentially, there is little prospect that human society will change the way that it behaves to meet important environmental concerns until most people believe that those changes are necessary. Only then will the decisions required achieve political popularity instead of the unpopularity that would greet them today. Achieving this change can only be through education; it cannot be through other forms of 'social engineering'.* [5]

He points out that this education can be informal, largely through the media, or formal, through organisations. The media has been helpful in informal education programmes on matters that affect the individual such as smoking, AIDS and drink-driving. The situation is likely to be very different when measures are advocated that restrict growth and would therefore affect profitability. Smith points out that formal education need not be restricted to our education system *per se* but can be through other organisation – religious, conservation, ethnic and social. As a leading educationalist he recommends that primary and perhaps secondary education should focus on the 'three Ps', poverty, population and pollution, but universities have a major responsibility to give fundamental consideration to the 'three Es', economics, ethics and ecology. He is thus talking about value systems and the relationship to nature which we have discussed here. His thoughts and conclusions on economics, ethics and ecology are similar to ours.

Sir David Smith describes his own attempts at environmental reform in Edinburgh University:

> *All activities of the University should be examined to explore whether an integrated 'environmental initiative' could be developed, partly analogous to the current 'enterprise initiative' but also extending into spheres other than teaching. An integrated environmental initiative would include not only teaching and research, but also the ways in which these link with environmental aspects of how the institution itself operates – 'institutional behaviour'.[6]*

This was written in December 1990 and we could comment that if one of the leading and most innovative universities in the world had reached this point in its thinking only in 1990 then there is a long way to travel for all of us. Indeed, it seems that the contemporary axiom of universities is whether money-making initiatives can be developed.

He divides his environmental priorities into first order and second order. First-order problems – poverty, population and pollution – survival of civilisation as we know it and second-order concerns which are deteriorations of the quality of the environment in which we live but which do not necessarily constitute a long-term threat. Thus global warming will be a first-order concern while some forms of wildlife preservation will be a second-order concern. It is our view that so-called second-order concerns are of paramount importance in developing community interest in the environment. Community education represents the only way forward in a democratic society if we are to survive, and the second order concerns – those about whales, koalas, threatened species and quality of air in a suburb are an important opening of the door and the mind for countless individuals. They provide an initiation, a recontact with nature, the environment and health which cannot be provided by the recycling of rubbish or the solar heater!

We conclude on an optimistic note. In today's universities some of the more mature, thinking students recognise the issues for themselves. We found the following statement in a student essay on environmental issues:

> *It must be painful to be an aware parent seeing the education system turn your child into another of society's drones, yet there are few choices, but some hope. There exists a handful of schools that focus upon self-development instead of mass production. However to sit around and hope is never terribly useful. It is more fruitful to work on the issue and the path will become apparent as we travel.*

Educational reform is essential for the greening of health. Health and the environment are tightly linked. We cannot hope to improve health and well-being without reforming our education system to replace or at least balance our consumer culture with personal, social and environmental values. War confers the most extreme example of ill-health, yet where in our education systems do we have reasoning and self-analysis to avoid it?

Attitudinal change

There have been many successful health education programs, including anti-smoking, anti-cancer and vaccination programs. In promoting ecological health we are aiming to change attitudes to the environment by means of education. We realise the impossibility of considering the behaviour of individuals without considering their circumstances. If people are poor and starving their first interests are likely to be survival, even if it is at the expense of the environment. People who are rich, but ignorant of the things we have written about, may equally be insensitive to the world's ecosystems. Moreover, there are those who are quite aware of the environmental problems we face yet act in an environmentally irresponsible way.

A few years ago there was a flurry of interest in environmental protection, with increased membership of environmental organisations and some initiatives by governments. String bags appeared in supermarkets, people recycled their plastic bags or brought sturdy ones to the shops. Our impression, from anecdotal evidence, is that these activities did not last for long. Most people get new plastic bags at every shopping trip, and without doubt the fundamental ways in which we live have not changed at all. We still drive cars as much, or take trips in aeroplanes, or eat foods grown thousands of kilometres from our homes, eat lots of meat, and generate vast streams of waste. In the light of the sure knowledge that doing these things is damaging to our environment, why haven't we changed fundamentally? How difficult would it be to do so? What are the roles of psychology and of sociology?

Humans do three basic things: we think, we feel and we behave. Each of these basic processes exists, on the whole, in connection with the others. For most of us living in the Western world, changing the way we live so that it is ecologically sustainable might involve attention to all three basic processes. One of the causes for our seemingly inexhaustible consumerism may be the positive feelings we get when we buy something or experience something new. It follows then that if we are to live differently, there will have to be a replacement for consumption, just as consumption took the place of ritual and religious worship.

This model of human behaviour and change emphasises fundamental human qualities or needs which have to be met. They include the need for novelty and stimulation, the need for relationships, and perhaps a need for spirituality, including what E.O. Wilson has called *biophilia*: a connectedness with the natural world.[7] The practical application of such a model is to enhance in humans those qualities that are consistent with ecological sustainability. So, argues Wilson, if we come to love nature and to value it emotionally we will also want to protect and nourish it. Movements such as deep ecology and environmental ethics give to nature some rights of existence similar to those enjoyed by humans. Whether such ideas have or will become dominant concepts is doubtful.

To a large degree the environmental impact of each person is small yet when aggregated it is very large, especially the impact of Western communities. Therein lies one set of problems which may make it difficult for us to change. The distance between our behaviour and the environmental damage we cause is by and large very great and involves innumerable steps. Thus if we go into a shop and purchase a book that has been printed overseas, we engage in an act that involves the destruction of trees, possible pollution from paper production, and production of greenhouse gases and other pollutants from local and international travel. Should we not read? Hardly. But there is an alternative: use a library. And yet many of us do not, largely because of the convenience of owning books. We do millions of things which make absolutely no sense in terms of ecological sustainability. How much of it could we change as individuals and how could we go about doing so? If enough people acted in environmentally benign ways there would be a massive effect on the production of goods and services. Why is it, then, that enough people do not do reasonable things to improve the environment? The answer to that question is complex and again involves the interaction of thinking, feeling and behaving in a particular context.

Consider the decision of a member of a two-car household to switch to a bicycle out of environmental and health concerns and to save money. Now while one rides happily, the other family members complain about the increased strain of having only one car. These strains include the extra burdens on whoever has the one car to transport children to and from various venues, and the increased rush in the mornings since one person may have to make two journeys. The consequences of those strains include increased tiredness, irritability and poor concentration, which are clearly adverse in terms of health. The convenience of having two vehicles unquestionably makes life better. While this might be a trivial example from someone who can afford the luxury of two cars, the story nevertheless does illustrate how the structure of our day-to-day lives affects how we are able to respond to simple issues. The example shows that making change in one area for sound reasons can have detrimental consequences in others. That may even be true in a very individualistic area – what you eat. To decide to become a vegetarian has a profound effect on shopping and cooking times.

The point of these discussions is to argue that unless one is effectively asocial it is almost impossible to make apparently positive environmental change without affecting other people especially those closest to you. Therefore environmental change at the individual level becomes one of discussion and debate among those people and sometimes views will be divergent and perhaps incommensurable. If such is the case, living with alternatives and accommodating other views and behaviours becomes a primary task. Whatever we desire to change will always be context dependent – if governments legislate for recycling then the question of individual preferences and the debates alluded to above become subordinated to what is in effect an imposed standard. A central question for ecology is how do we bring about the changes required to ensure sustainability and equity?

Perhaps the most cogent model for change independent of coercion is that of behavioural psychology. Allen Wheelis, in *How People Change*, reminds us that our identities are defined by our actions.[8] A thief is a thief because he or she thieves, a conservationist is a conservationist because he or she conserves. Having insight into and feeling for ecological issues means very little if not accompanied by relevant actions. At this time in our history we are due for and in need of a change in identity. For those of us in the West that change could be to take on an ecological perspective on life, to see life through ecological lenses. The same applies to the Third World, although the actions needed are very different.

Taming the economic juggernaut

We have argued that the present economic system is ecologically unsustainable and is responsible either directly or indirectly for the environmental perils that our wounded planet now faces. This theme has been a common one among ecologists and environmental philosophers. Yet it soon becomes very clear to any reader of this literature that very little has been written about what can be done to change this state of affairs. Much has been written about the nature of future, ecologically sustainable societies: they would be steady-state, zero-growth systems using renewable energy, practising recycling and with a high degree of local self-sufficiency, and community values would be encouraged to flourish in place of the religions of consumerism and materialism. Little, however, has been written about how to get there.

The question of the exact nature of the ecologically sustainable 'conserver society' has not been our focus of concern. We believe that the question of 'how to get there' is now the most crucial. Unless we begin constructive work towards a goal it is very unlikely that we shall get there. How can we tame the economic juggernaut before it destroys our health and the planet as we know it? That, in our opinion, is the most difficult and most important political question that can be answered today. So how can it be done?

We do not doubt that education will play an important role in the transition to a sustainable conserver society, otherwise it would have been pointless to have written this book. Education is a *necessary* condition for change, but is it enough, that is, is it sufficient? We think not. Capitalism has a vice-like grip upon our lives. In modern times, orthodox religion has come to occupy a back seat to the religion of consumerism which promises heaven on Earth *now*. This will be a grip which will be difficult break, especially within a relatively short time frame. Capitalism took about 500 years to develop into the fully grown monster which it now is; the transition to an alternative society is unlikely to be achieved in a short time.

Should we advocate revolution as Marxists have done? This, we believe, is even less satisfactory. Ignoring all the questions of the morality of revolution, Marxists were unable to create a sustainable 'communist' society on the basis of revolutionary practice. The achievement of ecologically sustainable societies across the Earth will require social and intellectual 'revolutions' greater than any that have occurred in human history. Revolution and warfare are unlikely to help us build the conserver society. Indeed the revolution to finish capitalism will be so bloody and dirty that it may push the planetary ecosystem to the point of ecological collapse anyway.

It seems that the conserver society will be adopted as a matter of historical necessity. Ultimately the instabilities of the capitalist system may force people to live simpler lives. Unemployment, urbanisation and increasing city housing prices, along with ethnic and racial tensions, crime and pollution will encourage many people in the West to move back to rural environments. Eventually, capitalism may tear itself apart as Marx predicted, but not for the reasons that Marx gave. Capitalism and communism are growth systems and growth cannot continue for ever. So ultimately capitalism will fall apart, just as communism has fallen apart. The system will become exhausted.

Given that growth causes significant environmental damage and adverse changes to the global ecosystem, it cannot continue forever. Therefore we must learn to live within limits or not live at all. We favour a *pluralist* strategy for social change. We support changes in education as well as local political activism to protect local environments. We support actions directed towards minimalising international trade, and developing local self-sufficiency in environmentally friendly ways. There is thus no reason to be apathetic and adopt a fatalistic attitude of 'letting it all happen'. For intellectuals it is especially important to challenge the status quo, which is still locked into the growth cult. It is especially important to undermine the philosophical underpinnings of this world view. That has been our key aim.

It is important that we offer readers at least some concrete proposals for action, rather than leave them with vague generalisations. We are opposed to internationalism which, tied to the growth of capitalism, has become a globally destructive force. What can specifically be done to break down internationalism? For a start nations – while they still have the power – need to re-regulate their economies. The situation of the values of currencies being determined by global market forces leaves a nation at the mercy of these global market forces. Foreign investment and the ability of trans-national corporations to move capital in and out of countries must also be controlled. At present countries are beholden to the major multinational companies. Countries are picked off one by one by the threat to jobs and withdrawal of investment. A 'trade union' of countries to counteract this power must evolve. If the United Nations cannot or will not take on this task, another democratic organisation must do so. All of these changes can be made; they existed in the past and with the political will, they can be implemented again.

Globalisation and free trade continue to expand with pre-dictable adverse effects on the environment and health. Each country has to remain on the conveyor belt or it is left behind. Means must be found to protect the environment and the health needs of indi-viduals within this globalised system. If subsidies and tariffs can be abolished by worldwide negotiations, it must be possible to negotiate soil conservation measures which all cereal growers must adopt before they can sell on the market. That way the cost would rise for all producers and no one would be disadvantaged.

One of the most important changes that needs to be addressed is the unstable nature of capitalist financial institutions. Here we are entering a complex field but also a field full of lies and deception of such magnitude that most people have accepted these lies as reality. Consider for example, money. Most people take money to mean paper money – US or Australian notes for example. In the case of Australia, in 1995 notes made up only six per cent of the money supply of 323 billion dollars, a paltry 19 billion dollars. However the Australian money supply increased by about 21 billion dollars in 1995. This increase in money is not an increase in money created by the Australian government's Reserve Bank, but merely a creation by the private banks. Indeed, it is believed that bank lending (including credit facilities like credit cards) is ten times the amount banks receive in deposits. Royal Commissions investigating the banking system in Australia, Britain, Canada and New Zealand have confirmed this. It is for this reason that Sir Josiah Stamp of the Bank of England said earlier this century:

> Banking was conceived in iniquity and born in sin ... Bankers own the Earth. Take it away from them but leave them the power to create money and, with a flick of the pen, they will create enough money to buy it back again ... Take this great power away from them and all the great fortunes like mine will disappear, and they ought to disappear, for then this would be a better and far happier world to live in ... but if you want to continue to be slaves of the Bankers and pay the cost of your own slavery, then continue to let Bankers create money and control credit.[9]

Abraham Lincoln said that creating and issuing money was not only the supreme prerogative of government, but its greatest creative opportunity. Throughout the West various economic reform groups – such as COMER (Committee on Monetary and Economic Reform, Canada) have been working on this program for nations to regain control of finance from the private banks. This is a long and painful battle because we live in an age of privatisation. The regaining of financial control by governments – who in turn must be subjected to strong democratic control – is a necessary step in the creation of sustainable societies. Ted Trainer, in *Towards a Sustainable Economy: The Need for Fundamental Change*, explains the process of credit creation in further detail as well as suggesting some interesting and radical reforms for the financial system.[10]

The regaining of financial control means just that – giving government ability to provide employment and complete tasks essential to all the community and to their future. Society has many tasks which will not make a profit. Within the sphere of scientific medicine, those services amenable to profit have been embraced by the private health sector and private hospitals have flourished on them. Left aside are the intellectually impaired, those with psychiatric illness, the old and the less well off – a shrinking public sector can look after them. Take a further step backwards from the needs of society and we find that environmental health measures are almost entirely dependent on finance from the public sector, they do not 'make' money and are regarded as burdensome on budgets; their research funding is a pittance compared to funding for 'scientific' medicine. Take a further step back and we see ecological health and its pitiful financial funding in today's budgets. Even overseas aid, which can do something to alleviate poverty and the environmental degradation it causes, is subject to pruning.

We must work for a reversal of economic policies so that the public sector is increased, enabling it to provide the infrastructure society needs for a healthy life. The main pillars of this infrastructure will not be roads and bridges, but environmental repair, community health, social structures and increased employment within these essential services. The financial developments proposed by COMER would allow this to happen by tackling basic problems with the system of financing infrastructure. It is not generally understood by people that banks create money (in the form of credit) when they make new loans: the money supply increases by the amount of the loan. When the loan is paid off the money supply decreases by the amount of the loan principal. This system is good for the banks as it essentially gives them a licence to print money. But it also creates instabilities because money is created out of nothing as a debt to the banking sector. Economic crashes occur because this system causes debt, and interest on debt grows by compound interest, much faster than money and income can grow. COMER has suggested a Sovereignty Loan Scheme whereby money is created by government at low interest rates for essential infrastructure. This is being pursued by states, cities and communities in Canada and the USA. In

Australia, environmental repair, despite degraded rivers and hundreds of thousands of acres of salinated land, has such a low priority that one government insisted on the sale of a public communications utility, Telstra, before repair could commence. What an appallingly poor level of commitment to the future. Such repair could commence at any time under COMER principles.

The rebirth of the public sector needs an attitudinal change. Does it matter if its 'efficiency' in economic rationalist terms is only 75 per cent of that of the private sector which motivates itself by profit? The 25 per cent 'slack' may in fact be spent on community involvement, education and consultation – no profit but a stronger society. It may be spent on more worker participation – again building a stronger and more caring society. In many fields, such as medicine, nursing, social work, the law and public utility management, there are still important pockets of commitment and service to the community. There are people who have withstood the manipulation and browbeating of some politicians because they believe in service, not just profit. Such public services provide strength to society, not only as a back-up for when capitalism collapses, to be called in as a 'new deal' by a Roosevelt, but to provide the everyday basic health and living needs for society.

In conclusion, economic reform is necessary for the greening of health. The global environment and therefore our health and well-being will be destroyed eventually by continued 'growth'. Already we are finding that economic rationalism has determined that budgetary indiscretions such as 'failed developments' shall be corrected by cuts in the health, education and environment budgets. These have become the 'non-essentials', the victims of the ideology of 'smaller government' rather than the life support systems for all humanity.

The community and politics

Reform will have to come by peaceful revolution against liberal democracy in the name of democracy. This will be a challenge requiring revision of the liberal view of citizenship, which is limited to exercising rights to vote, speak, worship, acquire property, have it protected and have a free trial. This system allows us to say and do anything so long as we observe the law and do not interfere with the rights of others. So we accept the economic system, engage in a self-centred aestheticism and private fulfilment, and allow the social engineer to improve the lot of everyone by piecemeal reform through pulling mechanical levers. We believe this liberal sensibility, with its poets on one side and engineers on the other, will disintegrate.

We believe that the challenge to the increasingly anti-democratic tendencies of the present institutions will come from local communities concerned with local issues. Local, decentralised and self-sufficient Green regions in different continents will not be taken in by the technocratic liberal facade. The technocratic world view will be opposed. The myth of the domination of nature and the worship of high technology will become transparent. Let us look at some mechanisms for change.

The initial opposition to globalisation may well arise in local communities because the global market aims to destroy their particular way of life based on caring for the land. Furthermore, 'place' means more than the land or natural environment; it is also a built environment, and we can care for the city because it is our home. As globalisation extends its tentacles, resistance to the juggernaut of the universal market by local communities in different regions will develop in different ways; in general they will unplug themselves from it, so that they can determine their own way of life. In Australia, as in other countries, some regions will try to become self-sufficient and Green whilst others become increasingly cosmopolitan and more embedded in the global market. There will be conflict. This will rupture the federal structure of nations and the break-up will foster change. Those regions, for example large parts of Africa, at present in chaos from the lack of social structures, may be well placed to develop new, localised, self-sufficient societies.

Our feeling is that the cosmopolitan regions will become increasingly extreme. They will continue to surf the global waves to make a fast buck anywhere, and will continue to see themselves as the crowning achievement of world civilisation. They will proclaim, after doing their cost-benefit analysis, that there is no desirable form of life or way of being-in-the-world other than their own. Everything else, such as the preservation of wilderness, is a dangerous, nostalgic romanticism.

The Green regions in Australia and other continents will protect their regional home on the grounds that humans can only become healthy if they have an ecologically sustainable urban form within a well cared for natural landscape. These Green regions will be centres for wilderness in Australia, and their thinkers will argue that the good life is the human life well lived, in contrast to being the smoothly functioning machine in the artificial world of a technopolis.

Changes in attitudes are occurring already in many democracies. Political systems are being questioned and rejected because it is realised that it is naive to expect change and reform of the state to come out of a political process that serves the interest of global capitalism. This rejection of political processes is seen most clearly in countries that do not have compulsory voting.

We believe that reform may come as follows: instead of making the state the primary focus of political action, as did the old Marxists, and then adapting citizenry to the state, a community or region will adopt the mode of everyday political activity appropriate to that place. Political being will come to draw its sustenance from our relationships among family, friends, church, neighbourhood, workplace, community, town and city. These enable us to act together and provide the moral guidelines to effect decisive change. Whatever their current limitations – and there are many – they can provide constructive forms of rejection of the global market. Regionalisation may be facilitated by the worldwide communication systems such as the Internet which will eat away at the tax base of national governments. This will lead to regions needing to organise their own necessities through voluntary work and personal contribution.

By working in the world economy and living in local cultures and communities, the new social movements may gain control over the decaying old industrial cities that did not make the transition to the new information society. No one else wants these cities. Already they are being discarded and abandoned by capitalism for greener pastures – literally. Maybe it will be in the city neighbourhoods that local control will be achieved first in the name of better quality of life, increased regional and local autonomy to prevent environmental damage, and citizens' participation in civil society. In Australia the regions linked to the Sydney, Melbourne and Canberra axis may go cosmopolitan and become more integrated into the circuits of the global market, whilst others, like South Australia, Tasmania and Aboriginal regions, will become community oriented, as they are increasingly by-passed by the global market and in economic decline.

Where then? The conflict between the different regions in a federal republic could develop into democratic local government and decentralised federalism. This would give the Green regions greater autonomy within the federation of regions, allowing their citizens to pursue their Green way of life. In Australia this would result in a minimal form of federal cooperation with other self-determining regions, or secession. The Green regions would be increasingly obliged to sustain themselves from their own regional economic activity, which they would control to ensure that it was sustainable.

An electronically interconnected and cooperative federation of regional governments could provide mutual support.

It is our view that liberal citizenship will have to be rejected to avert environmental calamity because of its emphasis on citizens using their freedom to promote their self-interest. It will be challenged by the need for common action to create an ecologically sustainable society. The failure of the global economy to overcome increasing unemployment and resolve the ecological crisis with high-tech fixes may provide the trigger for liberal citizenship being replaced. A new form of citizenship in civil society will be grounded in common ties, values and history; it will hold that eco-citizens should be in control of regulating the economy; it will aim at creating citizens who realise the goal of living a full human life, well within a conserver society. This democratic citizenship will invent the forms and practices of an ecological way of life, new identities for a people united in a semi-autonomous region, new concepts of common concerns, and new ways to systematically intervene in the running of a country. It will develop a certain idea of a political community – one centred on an ethical bond between citizens pursuing a common substantive purpose, even though they are engaged in many different enterprises. Such developments will offer a community-based support for a healthy society and environment.

This devolution of democracy to the community poses the following questions and problems. Regional governments in many countries have frequently displayed incompetence and have been beset with financial disasters. In Australia it is difficult to find a state government that has been financially competent while showing respect for democracy and citizens' rights. We believe this situation is a reflection of the present economic system rather than regionalism *per se*. A century of increasing regionalisation will create societies that will still suffer from the effects of global warming or the galloping market economy. It is difficult to see how this can be resolved. Finally, our truly democratic solutions to prevent global disaster may be arrested by one of our periodic descents into war, which will crush both market and Green societies.

In practical terms, what can each of us do? Some form of local government exists in most democratic countries and many individuals committed to community well-being have been part of such government. However, too often those involved for personal gain, the 'development' and real estate interests, have held sway. Local government needs those committed to health, the environment and social equity. It needs proper representation from these spheres as well as the unemployed, the retired and youth. Each of us can consider standing for election or serve on non-elected council committees. We can lobby for reform to bring to local government non-elected representation from community groups. Community interaction at the local level is an educational process. We can be involved in many different community organisations – environmental, charitable, social. It has been explained in this book how privatisation has excluded the community from health and public utilities. The community needs to reorganise into surveillance and pressure groups on such issues. All levels of society, including the unemployed, need to be attracted into such community activism.

Will political change affect our greening of health?

Let us consider health in its narrowest sense, the health services provided in Western societies. Services are evolving to provide expensive high-tech medicine for the few. They are reducing budgets, increasing privatisation, fostering a more powerful medical profession. This profession advocates predominance of interventional methods, such as prescription of expensive drugs, which supports a huge pharmaceutical industry, and the use of expensive surgical treatment, sometimes for cosmetic reasons. In such a system, health prevention, care of the aged, infirm or psychiatrically disturbed, and treatment of chronic illnesses, receives less attention. Currently there is a push worldwide for decentralised health systems, which aims to break the medical dominance of the doctors, but for the wrong reasons: the doctors are seen as the high spenders and the cause of large government health budgets. The new democratic process in a civil society would certainly change the present systems of individual health care but not for the same reasons.

Green medicine would reject the medical model of much illness. The concept of health is a product of political, social and economic conditions. The judgement that someone is ill is to some extent a value judgement and not about the symptoms of the malfunctioning machine of industrial civilisation. Therefore the values that doctors use in making their judgements about what constitutes illness are those of society; they are *our* values, not the values of the individual doctor.

It will become apparent that the doctor's exclusive power to grant or withhold to his or her patients the status their illness, will be deemed to be illegitimate on the grounds that much of the power that the doctor has to control people's health rightfully belongs to the people whose lives are controlled. Health services have become a political football with decisions made without significant input from the community. There is no equivalent of the conservation movement that tries to protect the environment. Health services, instead of becoming health-oriented, have become consumer-oriented, with clinics and hospitals providing a profit-oriented service run by investors who aim to accumulate wealth for themselves.

As long as we wish to retrieve our accident victims and treat our medical emergencies, high-tech medicine will be required in some centralised hospitals. These will require community support and understanding. Much of the remaining individual health care will return to community clinics which will feed into community resource structures. In these centres, medical dominance will be dismantled and the role of the doctor will change. At present doctors are integrated into the state and have become agents of social and political control even though they perhaps think of themselves purely as scientists. Their role in the community will become one of facilitators, preventers of ill-health, role models in teams, and their training for this will involve developing skills in relating to people, leadership and education.

A democratised civil society will win back from government and doctors' organisations a control over health and human well-being which properly belongs to the community. The power and responsibility for creating a society of healthy citizens lies with the citizens and not totally with professional organisations which have often used their clout to defend their power and privilege.

This brings us to the real need for this change – the global necessity. The broader health issues associated with an ecological crisis are far too important for a state to ignore. True political freedom requires the health market to be regulated so that the health system can enable citizens to realise a Green way of life in a republic. In Green medicine, citizens are involved in the value judgements about the criteria for illness or health, and the doctor must share with citizens the power to make these judgements.

---

1. David Orr, 'What is Education For?' *Trumpeter* 8:3 (1991), pp. 99–102.

2. Ibid.

3. Ian Lowe, *Our Universities Are Turning Us into the 'Ignorant Country'* (University of New South Wales Press, 1994).

4. David Smith, 'How Will It All End?' in *Health and the Environment*, ed. Bryan Cartledge (Oxford University Press, Oxford, 1994).

5. Ibid.

6. Ibid.

7. E.O. Wilson, *The Diversity of Life* (Belknar, Harvard, 1992).

8. Allen Wheelis, *How People Change* (Harper & Row, New York, 1975).

9. Quoted in Louis Evan, *Sous Le Signe de L'Abondance* (In this Age of Plenty; The Pilgrims of Saint Michael, Rangemont, Quebec, 1988), p. 96.

10. Ted Trainer, *Towards a Sustainable Economy: The Need for Fundamental Change* (Envirobooks, Sydney 1996).

CHAPTER 10

EVOLUTION, ECOLOGY AND HEALTH

The disturbance of ecology that leads to famine, flood and social disturbance seems remote to us in Western communities. But what about the diseases we experience? As Greg Easterbrook said in *A Moment on the Earth*: 'Disease is distressingly easy to overlook as an ecological issue yet it is the world's worst environmental problem by a wide margin.'[1]

Recently a young Aboriginal girl became ill with a fever and sore neck. The illness progressed quickly. She became unconscious and died of meningitis. This was an illness that would fit into the spectrum of environmental and ecological diseases. Meningitis is infectious, it spreads most easily in poor social conditions and it affects the malnourished. But in the same week 40 or 50 Australians, most of them young, were killed in motor vehicle accidents. Another 30 took their own lives. Many more died of heart attacks, strokes or cancers. Many people experienced non-fatal illnesses. Indeed, three out of four Australians will experience one or more illnesses or injuries in any two-week period. The most common of these are headache, the common cold, high blood pressure, injuries, skin diseases, dental problems, arthritis and asthma. How could some of these various causes of death and disability be related to ecology?

What is true health?

Firstly, we need to return to the discussion in Chapters 1 and 2. We began with a realisation that health is not just an absence of disease, it is influenced by our environment, our social conditions and relationships, and the health of planet Earth. We saw that ecology seeks to understand the distribution and abundance of life by examining the conditions in which species live, the resources they require and how they interact with each other and with their environments. Ecology means many things to many people: for some it is a science, for others it is a political tool with which to fashion a better planet for all forms of life. For others still it is one piece of a jigsaw which has to be fitted with other pieces, such as economic policy, to benefit humans as much as possible.

Health and what it means to be healthy are also many things to many people. The meaning of health to a banker in Sydney will be extraordinarily different from that of an indigenous Australian in the West Australian desert. Within our ideas of health we can include those needs that we believe create good health. For most of us these would be a long life, freedom from incurable disease (at least until we are old) and an ability to overcome dangers to our bodies and minds. We would also like to be able to repair any damage to our bodies or minds that might occur. Just as in a romantic view of ecology, like Darwin's view of life on the river bank, there is seen to be some fixed balance of nature, some of these concepts of health see humans as wanting and seeking an ideal state. Anything that disturbs this ideal state could be regarded as abnormal or a disease.

Human cultures have always sought to explain abnormality, whether by witchcraft, science or mythology. The dramatic successes of clinical medicine and public health have made these fields almost unrivalled explainers of health and illness. But they have their competitors. Naturopathy, iridology and a host of other 'alternative' medicines and therapies seek to explain what is meant by 'healthy' and to correct abnormality when it occurs. As we explained in Chapter 7, the idea of what is normal or abnormal in health is related to our cultural beliefs. This applies as much to scientific medicine and public health as it does to alternative therapies. Even within Western cultures the idea that a person is healthy, unhealthy or somewhere in between, depends on the perspective from which we look at that person. Our neighbour smokes 50 cigarettes a day and cannot walk up two flights of stairs without getting short of breath. Is he unhealthy? Or, is he unhealthy only when he gets a disease such as bronchitis or lung cancer? A public health practitioner might say that he is definitely unhealthy because he is exposing himself to toxic substances. We know from observing populations of smokers that lung damage occurs even if cancer or bronchitis does not occur. Public health would say that it is 'abnormal' for humans to smoke cigarettes and would undertake epidemiological investigations to show that this abnormal behaviour causes disease. As a result it urges smokers to quit and non-smokers not to start. A doctor dealing with an individual patient would focus on the symptoms produced by smoking cigarettes such as coughing and shortness of breath – because these are 'abnormal'.

So we see that both clinical medicine, which deals with individuals, and public health, which deals with populations, have some idea of what is normal and what is abnormal. We can carry this discussion further: scientists may discover a gene that predisposes to lung diseases caused by tobacco smoke. A quick and reliable test of whether one has this gene might then become available. Those who had the gene would know whether or not they would get lung diseases. Those who did not have the gene would know the opposite. Immediately our concept of what is healthy or unhealthy could be changed – smoking for some people would no longer be abnormal or a health hazard to them – though the 'passive' smoke they produced might still be a health hazard to others.

How evolution affects our health

Charles Darwin explained how species 'evolved' or changed with time.[2] He put forward the idea of 'natural selection', whereby hundreds of offspring might be produced by a tree, a flower, a frog or an insect, each varying slightly between themselves. Those offspring with slight advantages such as size or speed or colour to camouflage themselves from prey would survive to reproduce, so transmitting their advantage to the next generation. By this means species would 'evolve' or change.

Human beings are just as subject to the same major rules of evolution – which is related to many of today's common diseases and how ecology interacts with evolution. Evolution is concerned with populations; individuals are dispensable in the process. Paradoxically, individuals are also indispensable in that they transmit the genes. If they have healthy bodies and minds then there is an increased chance that they will transmit their genes. But it follows that once reproduction and nurturing is over there is little advantage in the health given to us by our genes and evolution.

In human evolution, natural selection (and survival of the fittest) has ensured that many diseases die out. The gene that causes congenital heart disease, 'a hole in the heart', would be rare because most affected persons would not survive to have children. The condition would remain rare. Another condition, arthritis, would remain common because the genes carrying it do not cause the joints to become diseased until later in life. Those with arthritis will have reproduced and nurtured their young before they become incapacitated. So 'natural selection' culls really bad genes, that is, the ones that create illness early in life.

We can study natural selection today in some of the world's populations. Nauru is an island of only 21 square kilometres in the South Pacific. It has a population of 9900. The previously large number of inhabitants with diabetes has greatly decreased over the past 20 to 30 years. Yet their dietary behaviour – eating too much fat and sugar – suggests that diabetes should still be common. It is thought that those who possessed genes that rendered their carriers susceptible to diabetes have gradually died out because the effects of those genes were so lethal. They were lethal because diabetic mothers and their babies were more likely to die in childbirth. That left a population composed of many more individuals with genes resistant to the effects of over-nourishment, which still persists on Nauru. But why then did the genes that cause diabetes exist? It is thought that the very genes that now harm Nauruans may have benefited them in times past when the supply of food was limited, because these genes favoured the efficient use of sugar and fat in times when food was scarce, before the coming of the 'Western' diet. These same genes may be a disadvantage when food is plentiful.

Let us place these events in Nauru in the context of ecology. Over centuries, the inhabitants of Nauru adapted to their environment and what it offered in terms of natural foods. Natural selection favoured those who could exist most successfully within the Nauruan web of life. Many individuals could not adapt to the sudden change from a sparse but nutritious diet to a Western diet and so died out. Thus the problem of diabetes, which we think of as a genetic disease requiring injection of insulin, could in theory at least be solved by returning to an 'evolutionary' diet. This has occurred among Australian Aboriginals: 'bush tucker', a diet to which their ancestors had adapted, has alleviated their diabetes.

It is likely that many of our common diseases reflect the sudden changes away from the ecology of our evolution, which took thousands of years. Heart disease, stroke, asthma and some cancers probably fall into this category. Whilst it is interesting to speculate how this may have occurred, it has no application to those who have these diseases here and now. But an understanding of ecology does allow us to list the ecological requirements for health that will have the greatest impact on the health of humanity as a whole. Indeed, understanding our evolution tells us about the likely range of circumstances in which good health can and cannot be expected. It gives us a way of making changes to our ecology to maximise our health. We can find these circumstances by examining our evolutionary heritage, by seeing how health has improved throughout history and by examining how and why health is better in some parts of the world than others. In this way we see what is 'normal' healthy living, just as a clinical doctor describes a 'normal' heart sound, or a public health doctor creates a range of 'normal' values for blood pressure or weight in populations. When we link normal healthy living to how we are – and need to be – part of ecology, we create 'normal' ecological living. Just as the physician uses 'normal' blood pressure to promote health and treat illness, we can use normal ecological living to do the same; we can take the necessary corrective action, as long as we have not irreparably damaged the ecosystems on which we depend.

Ecological health, then, is an extension of clinical medicine and public health for both individuals and populations. Ecological health recognises what is abnormal in an environment and tries to correct this. Delivering clinical care to the individual by fixing a broken leg or a public health measure such as vaccination against measles amounts to little if starvation, dehydration or lung disease supervene. In Western society the causes of ecological ill-health are less obvious and relate mostly to changes from our evolutionary lifestyle. The influence of diet has already been mentioned in relation to diabetes, but inadequate exercise, persistent consumption of chemical residues from pesticides and herbicides, drastic changes in social and work organisation and having children at sub-optimal times all have health implications. Some of these changes are correctable under our present social structures, some are not.

Breast cancer is the biggest killer of young women in the Western world. The risk of getting this cancer increases with a high fat diet and with alcohol consumption and when there is a family history of it. There is also an increased risk when the birth of the first child is delayed from the early twenties to the late twenties or beyond. The more menstrual cycles a woman has before giving birth the greater the risk. Because most women have shifted their age of giving birth to much later, this increased risk affects many more women than other risks such as a family history of the cancer. We attempt to solve the problem of breast cancer by X-ray mammography screening and by early surgery and other treatments. The only reason we do not advocate an evolutionary approach is because economics drives the way we live. Today motherhood in Western communities is a conscious decision which relates to career choice and economics, but this conscious decision should include a full knowledge of the health implications.

Each of us has a unique blueprint of genes from which all the structures and functions of our bodies are made and maintained. An inbuilt clock ticks away so that these structures and functions decline naturally with age. An inherited defect in a gene may give us a disease or a cancer that will cut short the slow progress of the clock. However a normal gene may become damaged because we are exposed to radiation or carcinogenic chemicals in the food we eat, and so we develop cancer. In this case the environment has stopped our clock. These are simple examples and are easy to understand. But the interactions of genes and environments are much more complex. In some parts of Africa, for instance, the possession of certain genes protects those children from devastating cerebral malaria; children without these particular genes are many times more likely to die of cerebral malaria. The gene causes a disorder of cells in the blood which in itself causes illness. This illness has not died out during evolution because the defective red cells protect against malaria.

Our biological clock, which has the potential to tick for 70 or 80 years, may be stopped by a variety of environmental causes ranging from accidents to starvation. The life-span of our forebears was on average only 30–40 years and was dictated by harsh environments and a lack of or an inability to use resources. When we now live to be 70 or 80 in Western communities it is because of the absence of circumstances that shorten lives. Such examples are common in the mammalian world. The life-span of domesticated animals is longer than that of their wild counterparts. When the Australian wombat is brought into captivity to be well-fed and sheltered and in clean surroundings, it becomes a healthier animal because internal and external parasites disappear and weight is gained. By living in environments in which we have successfully reduced our exposure to infectious diseases we have made ourselves healthier in the Western world. The opposite is the case in Third World countries and within some communities in the developed world, for example, some Aboriginal settlements in Australia. So improved health is absolutely dependent on modifying ecological conditions – it is not about living in harmony with the whole of the biological world. If it were, many of us would still have a very short life-span and suffer enormously. In some cases we have needed to destroy a species to improve health, as with the smallpox virus. So, by understanding how ecology affects health we can maximise the benefits of ecological living while minimising the harms. Such an understanding returns always to a consideration of conditions, resources and interactions, the ecological framework on which we can build an ecological health.

As we have seen in Chapter 4, disease can result from the interactions between humans and other species and between individual humans. Aggressive interactions between humans such as war and individual violence lead to ill-health. Most humans no longer have anything to fear from other large species. Snakes, sharks, tigers and bears kill humans but by far and away the biggest killers of humans are the infectious germs and parasites. We constantly interact with these species (see page 40). When we are exposed to a species we have not encountered before or have been exposed to infrequently we may suffer an infection. Often we alter the conditions and resources that determine the distribution and abundance of these germs. They will increase if their living space or habitat is increased. One way of doing this is to increase the number of humans in any given area. For example, once a human population reaches about a half a million the conditions and resources for the measles virus to flourish occur. Smaller populations do not have endemic measles. Therefore whether or not we are exposed to measles depends on the ecology of numbers and close interaction of people. Science through immunisation has found a way of minimising transmission of measles by effectively decreasing the number of individuals who can catch measles and transmit it to others. If we did not have immunisation we could consider ecological solutions to an epidemic such as spreading the population far and wide! For the many other worldwide diseases with no immunisation or vaccination and in some cases no adequate treatment, ecological measures are the only ones available. For the diseases discussed in Chapter 4, the measures would include avoidance of mosquito bites and clearance of stagnant water near to habitations, the use of shoes and sandals to prevent schistosomiasis, and the prevention of faecal contamination of ground water. Globalisation (see Chapter 6) is accompanied by an increase in human travel, which is neither ecologically nor medically sensible when epidemics are in progress. We are not advocating the abandonment of international travel; rather, a strategic appraisal of risks and how they can be prevented.

The tuberculosis organism is the single biggest killer of humans on our planet. It kills about three million people each year and about two billion are infected. Tuberculosis is an excellent example of an ecological disease. It is a bacterium which probably evolved in domesticated cattle and horses and was transmitted to man by close interactions with them. It is transmitted from human to human by coughing. Poor nutrition and a weakened immune system (found in destitute alcoholics and impoverished people anywhere) allows the bacterium to cause a devastating illness. The situation is further complicated by the emergence of strains of tuberculosis that are resistant to previously effective antibiotics. Ecology and evolution of the bacillus have benefited the tuberculosis bacillus at the expense of humans; by downgrading previously effective treatment programs and by inefficiently using antibiotics, we have allowed some strains of tuberculosis to become dominant at the expense of others. Those bacilli that multiplied were the ones that were not killed by the inefficient treatments. Add the effects of social decline and we have a recipe for a public health problem of great magnitude. Conditions, resources and interactions have acted together to create a new scenario of disease. The tragedy lies in their predictability from ecological first principles: evolution of microbes can occur quickly; overcrowding and poor nutrition create conditions in which tuberculosis can flourish and be transmitted easily. The solution is to change the conditions, resources and interactions – by investing in creating and maintaining an ecological environment – so that it is difficult for tuberculosis to flourish.

Mental health and ecology

Mental ill-health can lead to suicide, violence and physical disability. It is a major problem in our Western communities. How could mental health be linked with ecology? It is linked in many ways. Humans identify with their natural surroundings; observe how many of us go to the beach or country for holidays, how many of us engage in recreation outdoors and if we cannot, how many of us watch wildlife documentaries. Such observations have led some commentators to say we need the natural world to be mentally healthy. Whether that is strictly true is not the point. The point is that many humans derive solace and pleasure by observing and being part of the natural world. It follows that if this is taken away or withheld or damaged, ill-health could result. Certainly, this is so with indigenous peoples. Further, we could argue that for those who have become accustomed to living without the natural world, that revisiting or becoming acquainted with it could lead to better health. Author Theodore Roszak says that we live in ecological insanity and that the therapist and the ecologist offer us a common agenda, which would benefit individuals and the planet.[3] That agenda is: scale down, slow down, democratise, decentralise. These, he says, are ecological goals that can heal the psyche and psychological values that can heal the planet.

Further, because ecology is as much about how we interact with each other as it is about how we interact with other species, our emotional development and personality are intrinsically ecological. Is it ecologically normal for children to be cared for by non-relatives from just a few weeks of age? What effects could such deviations from being raised in small closely connected family and community groups have on development of mental health? Like cigarette smoking, which exerts its effects years after smoking was started, our child-care practices could have a delayed effect on personality and lifelong mental health. Why are so many of our young people showing signs of depression and despair? Perhaps because of the hedonistic consumer rush of our competitive society, with things being more important than people. Some might be angry at such speculations being raised and we agree they are speculations. But they are not irrational or unreasonable and are centred on the idea that there are normal social conditions, ones that evolved over centuries. We agree that the isolation of women, and the situation of motherhood being valued less than commercial work, should not be tolerated. But we do not think the answer lies in driving ourselves further and further away from our evolutionary design.

We should also remember that evolution is a very slow process of change. Humans may not have the adaptive processes to deal with the increasingly rapid changes in society and work conditions imposed upon them today. We know that for many individuals a change in home or city can result in major stress leading to anxiety, depression or even physical illness. The attachment to place is broken and grieving occurs. Yet today's economic system demands recurrent retraining and change of jobs in any one lifetime. The natural continuity of life is broken. Economic competitiveness, leading to demanding educational curricula, imposes stress and indeed suicide on children in countries like Japan, Australia and the USA. All these phenomena may represent demands running ahead of our evolutionary capabilities.

We are unlikely to support our natural world until we develop a robust mental health which arises out of healthy interactions with our fellow humans. When we come to realise that our well-being derives largely from positive interactions with family, friends, neighbours and colleagues, we then begin to realise that many of the mental health substitutes of the West – drugs and alcohol, excessive work, entertainment, and so forth – are not needed quite so much. When substitutes such as consumerism damage the environment then it follows that less demand for them is likely to reduce ecological damage. Perhaps the most important requirement for good mental health of populations is to be employed in the widest sense. That does not necessarily mean a traditional job, but any useful, productive activity that occupies time, provides social interaction and satisfies the creative desires of humans. We have reduced some de-humanising work practices in the West. We now need to think about work with ecology in mind. We must ensure that the structure of work fits with our biological requirement for the care of children, the continuity of work, and the opportunity for adequate rest, relaxation and recuperation. These needs are ecological and evolutionary.

Economics and health

An ecological perspective on health, then, leads us to consider what is most important here and now for health, particularly of populations. It leads us to question processes that may contribute to ill-health and to rank them in order of importance. Our first obligation should be to ensure basic sanitary and hygiene conditions and that food, water and social resources are available. Other 'basics' include shelter, relative freedom from toxic substances, social support, and freedoms.

An ecological perspective takes our primary focus away from attempting to 'explain' and modify the causes and consequences of individual diseases. It refocuses our attention on satisfying human needs within an ecological framework. The resilience and resistance of ecosystems to damage is an important consideration. In an ideal ecosystem human needs would be met locally with as little ecological disruption as possible. Perhaps there is little ecological sense in those of us who live in the colder south of Australia eating bananas, which are grown in the warm north and transported thousands of kilometres, wasting energy and contributing greenhouse gases. At the moment such considerations would be regarded as irrelevant because as long as there is a market for bananas in the cold south it will be filled. That is, economics is given priority over ecology. What we need is economics integrated with ecology. At present, the 'market economy' would explain that jobs are created to give the southerners bananas. The grower, the transporter and the retailer benefit. Furthermore, it is a free market and we have a right to eat bananas. The ecologist will not deny the right to eat them but will point out that the massive monoculture of bananas has an environmental and human cost in toxic sprays, the transport which causes the emission of greenhouse gases and the use of roads which are heavily subsidised. The true cost of bananas is really six times the shop price! Would it not be more 'rational' to eat more locally grown stored apples in the cold south? Their production also produces jobs. We could still choose to eat bananas but at the true price. 'Rational' economics has bred irrational behaviour – witness the worldwide circulation of orange juice and its importation into countries that already produce orange juice (see Chapter 6).

Ecological health

Our modern way of life is so connected to economics and politics, it is difficult to think of ecological health without radical changes to them. It is hard to imagine a hamburger outlet distributing ecologically healthy meals rather than packaged fat, sugar and salt. Today, when finance dominates economics and making a dollar takes precedence over making better lives, it is hard to imagine the development of policies that would facilitate a better pattern of reproduction and care of children. And, as discussed in Chapter 4, it is difficult to imagine that the reining in of global warming will take precedence over our global economic machine. If we follow the line of reasoning developed in Chapter 6 we will regard health, education and shelter as essentials and not treat them as dispensable, to be funded only after the economic rationalists have had their way. On this basis we should ask what an ecologically healthy population or community would be like, and then fashion an economic system around it.

Focusing on ecology and asking such questions leads us immediately not only to greenhouse reduction strategies but also to individual and community decisions about what is normal ecologically, just as we have concepts about what is normal in regard to health. We argue here that health and ecology would then be integrated so that changes that result in improved or rehabilitated ecosystems will at the same time maintain or improve health.

We are not being naively utopian. All living things destroy and refashion their environments to some degree. This is a biological fact of life from which we cannot escape. But there are limits to how far we can go in affecting local and global ecosystems to meet our needs before the costs outweigh the benefits. This issue was recognised in Agenda 21, the main document resulting from the Rio Earth Summit which we discussed in Chapter 1. It was explained that health ultimately depends on our ability to manage the interaction between the physical, spiritual, biological, economic and social environment. Sound development is impossible without a healthy population and the lack of development adversely affects the health of many people. By 'development', Agenda 21 means the provision of requirements essential for health and well-being, the items we listed previously. Development in its broadest sense does not necessarily mean Jeep Cherokees, or Ok Tedi mines, or long-line fishing or wholesale forest clearing.

It is unfortunate that many of us in the West with interests in health and ecology have lost sight of the really big ecological problems that affect the health of most of the inhabitants of Earth. Greg Easterbrook is rightly critical of some of the outcomes of the Rio Summit when he observes that:

> There is something faintly indecent about the world's heads of state gathering as they did in Rio, to bestow many tens of billions of dollars on the greenhouse effect, a speculative concern, while lifting not a finger to assist 7.8 million children dead each year from drinking infected water and breathing dung smoke. Yet this ordering of priorities is in sync with contemporary environmental doctrine. A little reported part of Rio was a proposed agreement by which the First World would increase environmental aid to the developing world, for purposes such as water sanitation. Western

nations ended up rejecting this proposal pleading, We'd love to help but we just committed ourselves to big investments in fighting the greenhouse menace.[4]

It is not as if the world could not solve this problem. It managed to eradicate smallpox. The answers are known: the problem lies not in science or technology but in its application, which is fundamentally a socio-political and ethical problem.

Easterbrook is not arguing that greenhouse warming will not cause problems or that we should neglect it, but rather that ecological problems are likely to be much greater if we do not assist all populations on the planet to live in their environments in ecologically sensible ways.

It is a sobering thought that most of the actions of the authors and of countries such as Australia are unnecessary for achieving health. This is illustrated by the apparent paradox that some parts of the developing world such as Kerala in India and Sri Lanka, where income and resource availability is a fraction of that in Australia, life expectancy is only a little less than in Australia. What seems to be important in these regions is the equitable distribution of basic resources. This equity is ensured at least in part by political systems in which the power of women is considerable. Once we have the provision of food, hygiene and basic health care, personal lifestyle has the most important impact on health. This leads to a further paradox, which illustrates the importance of using resources in an ecologically optimal way: the excessive calorie intake in the West is a major cause of ill-health and premature death.

Our concept of ecological health is substantially different from the dominant concepts of health today. These give priority to treatment, to readily measurable and narrowly defined outcomes, and to a rhetoric of health promotion and illness prevention without a backbone of ecological social change. The status quo is maintained because it is tied to acquisition of power and resources. As we have seen this is highly mechanistic. We are willing to invest large sums in research projects which by any standard are of a high calibre but are of questionable relevance. As we have said earlier, Australia's premier health research body invested ten times as much money into a few narrow laboratory areas in 1993 as it did into public health and prevention. Few people question such disproportions and even fewer listen. But we are in a new era now. An era in which ecological medicine has more to offer us than further developing technological medicine. But ecological medicine requires a much broader and deeper politics than technological medicine. We are not talking about the politics of introducing screening for breast cancer or the politics of developing and introducing a new drug or indeed the politics of gender-related health. Rather, we are asking questions about the need for and means of making substantial changes in the way we live as individuals, as communities and as institutions. These will not be favoured by present-day economics or power brokers.

These substantial changes involve giving priority to a human ecology integrated with concepts of justice and fairness. The building blocks of such an ecology have been known for a long time and by many peoples ancient and modern. They encompass the beliefs about health held by indigenous peoples and the changes in economics and values in political structures discussed in Chapters 1 and 9.

Our argument for the necessity of incorporating an ecological dimension to health has examined population health separately from ecological health, so that we are able to understand the complexities of this aspect of health, which has local and global dimensions. Global cooperation and solutions will be required. This seems a daunting task when we see the communities of some of the poorest countries of the world tearing each other apart. A stronger, not weaker, United Nations will be required. National self-interest, tribalism and short-term self-interest will make it very hard to solve pressing problems. As discussed in Chapter 9, a generation of reform in education may be required before the severity of the problems in the West and the Third World are recognised. Complex problems require comprehensive solutions. By contrast, population health and individual health are much narrower in scope. For example, it is well known that we can develop drugs to ameliorate diabetes and asthma but that we have much less success in convincing people to take them appropriately. That is because humans are complex and are influenced by many things. Unless we pay attention to these complexities we can have all the bright ideas in the world but they will not come to much good. If we cannot treat diabetes successfully because of individual behaviour, imagine how much harder it will be to cooperate to change many behaviours that we unthinkingly regard as normal now. Ministers of health and environment might be sympathetic to the ideas presented in this book but in our current *modus vivendi* they are subordinate to their colleagues in economics and finance departments.

In Chapter 5 we have seen that during famine, impoverished environments lead to impoverished lives, which in turn lead to further damage to the environment and to ecological systems eventually becoming unsustainable. Our political systems and economic systems are apparently incapable of dealing with these situations. Millions can be undernourished and starving while food mountains remain unused or are discarded in developed countries. The free-market concept of living is in a major way responsible for these anomalies. Primarily it has been based on our dominance of nature, on a war against infectious disease which must be won by eliminating micro-organisms, and on the taming of environments and their use for our health and well-being regardless of the costs. Health has had to fit into this scheme of thinking. But the evolutionary view of humans we present here tells us that humans developed within the Earth's ecosystems and that our biological heritage has a much longer history and maturation than our relatively brief culture. To see health as ecological involves a cultural maturation that would rival the

Renaissance, and which, we hope, will see us emerge from the dark age of ecological damage to an enlightened, ecologically centred interest in ourselves and in other life. This maturation is not necessarily technological.

Population health depends on new technologies such as vaccines, cancer screening programs and health promotion. Ecological health will not be delivered by such approaches because it does not arise out of focusing on specific diseases, behaviours or solutions and as such cannot rely on technology to deal with them, regardless of whether that technology is a biochemical test or a media campaign. We have every reason to be cautious about relying on technological solutions. While technology has undoubtedly been helpful it cannot solve problems that arise out of complex encounters of humans between themselves and with other life. Our ecological health problems are fundamentally about such encounters. Nor can technology solve problems that are not easily localised and have many causes. It can only be a part of a broader and deeper problem-solving mechanism. Thus environmental problems such as global warming, deforestation, decline in biodiversity and increased interactions between all forms of life as a result of global transport cannot be solved without global cooperative efforts. There will be significant changes in ecosystems and for those populations that have no alternative sources of resources or that are unable to cope with climate change, the health consequences will be severe. Predicting those changes in detail and developing the social and organisational skills to respond to them remain great challenges for us.

Not only are housing, nutrition, individual rights and democracy important to our health and well-being, but the nature of our relationship with the land and with its plants and animals is also important. This idea challenges the assumption in our liberal culture that sees ecology as 'out there', and separate from us. We hope we have shown that it must be seen as 'in here', a part of us. We cannot be healthy, either as individuals or populations, without this realisation.

Our requirements for good mental and physical health are surprisingly few and do not demand many resources. We are, after all, fairly ordinary mammals in terms of our basic biological and social requirements. Where we are extraordinary is in our cultural requirements, which seem to know no bounds. Our apparently insatiable desire for knowledge, power, possessions and novelty is pursued by affluent nations with great vigour.

Ecological concepts of health build upon, rather than discard, previous concepts. We recognise that good health includes our relationships with other living things as well as how we organise ourselves socially. This recognition has practical and ethical implications for all aspects of our lives. We begin to see that we have a debt to other groups of humans and to environments that needs acknowledgement, apology and restitution. We begin to see that we need sensible, useful and sensitive rules about how to live. In these ways we will become functional communities. If we do not undertake these tasks we will continue to be dysfunctional as a species to the detriment of ourselves and our ecological communities.

A utopian ecological future cannot eliminate the risk of disease, but improved hygiene, adequate water supply and better social integration will certainly decrease the risks. Fewer deaths from over-consumption of resources will occur. People will on average live to be older and die from age-related conditions due to the natural decline in function of their organs. Mental health will be better because we have decided to invest heavily in evolutionary ecological approaches to social systems. That will mean care of the young will be uppermost in our thinking. Systems will be developed that adequately monitor mental health at the level of populations. Interventions to improve mental health will involve whole communities and political systems and will radically transform work and family life. Deaths and illness from exposure to toxins such as alcohol and tobacco will decline because mental health will improve and because a new democratic order will see them being unacceptable.

1. Greg Easterbrook, *A Moment on the Earth* (Penguin, New York, 1995).

2. Charles Darwin, *The Origin of Species by Means of Natural Selection* (1859).

3. Theodore Roszak, *The Voice of the Earth* (Simon & Schuster, New York, 1992).

4. Greg Easterbrook, op. cit.

CHAPTER 11

HOPE FOR THE FUTURE

A report to government

Now we return to our introduction, which described the death of a draft report commissioned by the Australian National Health and Medical Research Council on principles for an ecologically sustainable basis of health. The executive summary of the draft report pointed out that human well-being depends ultimately on the well-being of the Earth's ecosystems and an ecological approach to health is based on the recognition of the interconnectedness of all aspects of the ecological systems in which people live. Just as all parts and systems within the human body are interconnected (something that health professionals take for granted), in ecosystems too, everything is connected to everything else. This interconnectedness extends across the entire mantle of the Earth. The use of ozone-depleting substances and the emission of greenhouse gases have potential effects on the ecological processes on which health depends, both in Australia and across the world. The report recognised that true health was a very wide subject and indeed detailed such matters as the statements from the Ottawa Charter for Health Promotion (1986), which had indicated that the fundamental conditions and resources for health are peace, shelter, education, food, income, a stable ecosystem, sustainable resources, social justice and equity. They were covered by our report to the National Health and Medical Research Council. It is likely that bureaucracy and our colleagues in medical science could not accept the far-reaching implications of our statements. Why could they not cope?

Government and bureaucracy are compartmentalised. Government departments of mining and resources, energy, urban planning, the environment, and social services are separate and competitive. Bureaucracy cannot cope with the complexity engendered by these departments having a significant effect on health. The position of the bureaucrat is that such situations are unmanageable. Yet we see, every day, the health implications of decisions made in all government departments, implications that support our belief that true health involves both our ecosystems and our social systems.

We recognise that our definitions in relation to environmental and ecological health can be expanded further into the social realm. For example, in some states of Australia we have recently seen the introduction of a considerable number of poker machines. This army of machines stands in hundreds of pubs, clubs and hotels. Governments saw the opportunity to raise additional revenue and put out the usual statements about the number of jobs being increased. Within months there was a reduction in money being donated to charities which provide important food, clothing and shelter for the needy. Small shopkeepers noted a marked and sustained reduction in their takings. Thousands of those with little money to spend were nevertheless gambling it in the vain hope that winning would allow them to escape from their circumstances. Those in the health services warned of the dangers of an increase in family break-up, psychological problems and malnutrition in those families affected by addiction to gambling. In one state in Australia the Medical Association made a statement that not enough money was being spent on food and that children were suffering because of poker machines. An eminent politician with a finance portfolio responded by saying: 'What has it got to do with the Medical Association?' Hopefully the reader has understood our messages and we need comment no further on this political statement!

Every decision taken by society has an implication for

health and well-being. In particular, the effects of decisions relating to balancing budgets and moving capital affect the health and well-being of those who are vulnerable. At the end of the day those decisions leading to environmental degradation will affect the under-privileged first.

There was nothing revolutionary in our draft report to the Australian National Health and Medical Research Council. The discussions between the contributors to the report led to concluding statements that an adequate diet of healthy and nutritious food is needed for all people and that a sustainable food supply depends on maintenance of ecological processes. An adequate supply of clean water for drinking and sanitation is essential to health. Again, sustainable water supplies for people and ecosystems depend on the maintenance of ecological processes. The concentrations of greenhouse gases in the atmosphere must be stabilised because all people and the ecosystems on which human health depends need a stable climate. We drew attention to the need for all living things to have the protection of the ozone layer from harmful wavelengths of solar radiation. As well as affecting human health directly, ozone-depleting substances also affect other species and ecological processes on which human health depends. We stressed that all people need shelter and community. A growing proportion of the world's population is living in cities and the ability of cities to support healthy populations sustainably depends on maintenance of outside ecological systems. The health of all people depends on ecosystems, therefore, both people and ecosystems should be protected from exposure to harmful levels of poisons, harmful germs and ionising radiation. We drew attention to the need to stabilise human population to ensure a decent quality of life for all people and to ensure the continuity of the ecological processes that sustain life on the planet. Fossil-fuel-based energy depletes non-renewable resources and adversely affects ecological processes, which in turn affect human health. Ecologically sustainable use of energy requires the development of alternatives to fossil fuels. Equitable access to the Earth's natural resources within and between generations requires careful use of non-renewable resources and ecologically sustainable approaches to the extraction and processing of all resources. We also emphasised that human health depends on the maintenance of ecological systems and that biodiversity is essential to the basic operation of the Earth's ecosystems and so is a fundamental require- ment for health. Ecological sustainability is fundamental to the maintenance of psychological and social well-being.

We have explained these requirements for health, empha- sising ecology and impediments such as the economic juggernaut and growth of population. We go further than scientific exposition of our troubles and needs. We introduce the need for value systems as essential for our health. We have listened to the plea of Aitlon Krenak, who was expressing thoughts and needs common to all the remaining indigenous peoples of the world:

> *I am here as the son of a small nation, the Krenak Indian Nation. We live in the valley of the Rio Doce. When the government took our land in the valley of Rio Doce, they wanted to give us another place somewhere else. But the state, the government will never understand that we do not have another place to go. The only possible place for the Krenak people to live and to re-establish our existence, to speak to our Gods, to speak to our nature, to weave our lives is where our God created us.*
>
> *We can no longer see the planet that we live on as if it were a chessboard where people just move things around. We cannot consider the planet as something isolated from the cosmic. Respect our place of living, do not degrade our living conditions, respect this life.*[1]

Who will lead us to health and well-being?

Individuals feel impotent when they look at our democratic political systems and ask what is being done by our leaders. It has been interesting to watch the election processes in recent years in democratic countries like Japan, United States, Spain and Australia. The picture that comes into one's mind is of a group of earnest political leaders engrossed in a game of snakes and ladders, oblivious to the bushfire that is rapidly bearing down upon their unprotected building. In all these elections there has been little of relevance said about the major threats we face and the future of the human race. Each election is concerned with ideology, with a rhetoric of greed, with cutting up the cake and who is to get most, and with the performance of each country as it competes with others. The overwhelming considerations are economic.

It would be interesting for the reader to consider a list of countries of the world and to ask which of them will offer leadership, not because they have economic or military power but because they are able by force of argument, reason and inspiration to offer a vision of the future which others could follow.

The reader may turn first to the major 'powers'. One of them, the USSR, has recently disintegrated into political, environmental and economic chaos, the other, the USA, a huge military power, is partially engrossed in its internal consumerism, the film Oscars and racial division. Vice-President Al Gore, who would understand our themes and viewpoint, has written a book on the environment which is a major contribution. Yet he admits that even in his own country, his impact against the surge of consumerism has not been great. In *Earth in the Balance* he describes how he ran for the presidency in 1987.[2] He focused on global warming, ozone depletion and the ailing global environment. Scarcely a word hit the press – because the media was unwilling to focus on these issues. The media are perhaps the main impediment to global reform. Nevertheless, Al Gore persists with his environmental concerns through the President's Council on Sustainable Development. The other major powers of the world, such as China and India (major because of their military power and huge populations), are engrossed in their internal problems, which include feeding their populations.

Let us now look at the smaller countries for the leadership and inspiration that might come from them. There might be only half a dozen names that might inspire you. Germany has the potential to assist with the process of reform for it has demonstrated its leadership skills in Europe on the economic front since the Second World War. Furthermore, it has a political system that allows the voice of ecology to be heard in the halls of power. Sweden has offered innovation and leadership in the areas of social and industrial reform and has been helped in these tasks by avoiding war. So too has Switzerland, which continues to foster the work of several international institutions. Within the under-developed continents there are a few lights on the horizon – models of self-sufficiency which are very important ecologically, as in Eritrea.

Then we must mention Australia. This is an internally stable country with perhaps the most successful multi-racial society in the world. Its gross national product is higher than most but the true standard of living, when we take into account all the environmental and social attributes, is possibly the highest in the world. It has resources and could be self-sufficient. Yet it is suffering from the disease that has infected all the small countries: a preoccupation with globalisation and in competing for a slice of the international cake. Even Australia has not stopped to ask itself what it wants for its citizens and their well-being. Australia runs its economy with an apparently permanent trade deficit. Its consumer society wants more than it is prepared to produce itself. Its forests are woodchipped to help the balance of payments pay for imports of consumer goods that are not needed. Australians 'need' their expensive imported cars just like some of the world's indigenous people needed the brightly coloured beads for which they 'exchanged' their inheritance. Australia, of all the countries we have considered, could offer leadership and be a voice of the future, but its present policies and attitudes are an embarrassment. Therein lies the authors' frustration.

We must lead ourselves

How are we to live and how are you to live? What must we do to achieve optimal health while we sustain the Earth and its human and non-human inhabitants?

The answers to these questions are not immediately apparent. Peter Singer urges us to live an 'ethical' life in which decisions about day-to-day living include broader concerns, not just selfish ones.[3] Theodore Roszak wants us to develop our innate – albeit latent or repressed – ecological ego.[4] E.O. Wilson wants us to develop our love of nature.[5] Sir David Smith thinks we need to re-orient education.[6] So does David Orr.[7] Perhaps all of them think that there is little chance of substantial change until a majority of citizens think and act more seriously about environmental matters.

Others, such as Robyn Eckersley, are most concerned about forms of political organisation.[8] It is often said that politicians are followers rather than leaders. It is our view that the associations between party politics, the media and business provide an unfortunate leadership of liberal democracies. The subtlety of their actions belies their enormous power. Living in a democracy like ours is meant to provide great individual freedoms. But it is the paradox of liberalism that the hard-won freedom from the shackles of poverty, tyranny and underdevelopment has been lost to new shackles of inequity, covert tyranny and damaged relationships with each other and the Earth.

Those of us who inhabit the affluent world bear the greatest responsibility for refashioning the ways in which we live. As individuals we consume much more than we need and we pollute more. Therefore we have the greatest scope for decreasing our consumption and sharing our resources, knowledge and skills with the less well-off in the world.

How could we go about changing the way we live as individuals and how far do we need to go? Indeed, how does any person go about changing a way of living which is thoroughly imbued in them? For the health and ecological matters that concern us here, the range of behaviours and outcomes is enormous, as is the complexity. For instance, if each of us drove a car to work rather than ride a bicycle, the outcome in terms of pollutants and energy use would be individually small. However, when millions of people do so, there is a worldwide environmental problem, even if its major effects may be 30 years in coming.

Why not appeal to the sense of responsibility of individual citizens? We can and do so regularly – there is no shortage of information in books and other media informing us about health and environmental problems and what to do about them. Except for the most committed there is probably little incentive or reason to change individual behaviour by radically reducing motor vehicle use, by becoming a minimal meat eater, or by sharing a job. Promoting behaviour change may benefit from an understanding of human psychology. What proportion of the population will change just because they have been given information? What proportion will need to be motivated to change? These are largely unanswered questions in regard to environmental matters. We do know that despite widespread stated concerns about health and the environment, humans in countries such as the USA and Australia live in fundamentally unhealthy and anti-ecological ways. In countries such as Australia, particularly among the young, concern about the environment ranks highly among other political concerns. But so long as the power-brokers make it difficult for the majority to change, change is only likely to come when this power system is replaced.

Private transport increases the greenhouse effect. When a liberal government downgrades public transport in favour of the individual freedoms of the private car then it is difficult for people to give up using their cars. If they did there would be an impact on the manufacture, sale and service of cars. Many workers might suffer. This result might explain why the environmental concerns of contemporary political parties (with a few exceptions) are largely cosmetic rhetoric. The massive development of infrastructures that support the consumer lifestyle means that change will be difficult; it will require great sensitivity and skilful reorganisation. That task is perhaps easier than we might think. Most of us do what we do because of habit – and habits can be changed under some circumstances.

It is our opinion that the closer we get to some steady-state situation, both ecologically and economically, then the more likely we are to equitably satisfy the needs of humans and other species. Environmentalists are concerned about future generations. But it is a mistake to have a program for change based solely on future effects, which would have only a small psychological relevance to us today and allow politicians and others to deflect necessary actions onto agendas for the next century. We have to explain why today's actions are unacceptable in terms of ecological sustainability. The politician says that yes we should live in an ecologically sustainable way, but at the same time promotes practices such as export agriculture and the globalisation of economies, which are examples par excellence of ecologically unsustainable and potentially destructive behaviours. Unless we review all our present-day behaviours and practices from a perspective of sustainability we will never have any meaningful and attainable goals.

The author who is a medical practitioner realises that we have a greater chance of stopping a patient smoking if we focus on the damage to health that is occurring now instead of quoting statistics about the chance of lung cancer. Campaigns and education involving the individual do have some success; those involving the environment often do not directly involve individuals and there is rarely an immediate benefit. With this in mind we will need to demonstrate such benefits, both tangible and intangible. The enormity of the task can depress us and lead to feelings of futility. But we should not underestimate the power of small, individual decisions. Consumers do have power and sometimes power that acts against the dominant ideologies and power systems. In recent years we have seen this power exerted by the thousands of individuals who rejected the wearing of animal furs. Similarly, the suffering of animals in the testing and development of cosmetics has affected the consumption patterns of many people.

Humans are emotional creatures and feeling is paramount in everything we do. We are emotionally developed long before we are cognitively developed, which is one factor in drug addiction: the feeling satisfies inherent and intrinsic needs. Alcohol is not widely found in nature yet five to ten per cent of Western populations are addicted to it. If you ask alcoholics why, most will eventually get around to talking about how they feel. The same is true of other behaviours. We feel good when we buy things, when we travel, when we solve a problem, when we read novels, watch movies or eat fine foods. More than any other mammals we crave novelty, the legacy of our big brains. Novelty makes us feel good. This feature of humans sets us up for manipulation. Our behaviours can be channelled by marketeers who skilfully usurp our feelings. We are addicted consumers of fashion or pop music or academic life or conservation but are only temporarily satisfied when we acquire the thing we desire. We then want more varied experiences and more of them. This is the seed of environmental destruction which will lead to ill-health. To recognise, discuss and act on these insights might enable us to regain control of our manipulated feelings. We need to have ways to maintain our feelings based on thoughts and actions that are consistent with the ecological ethic. In the past the desire for novelty was met by day-to-day life, by an appreciation of seasons, by intense socialisation and by ritual. Redevelopment of some of these lost sources of satisfaction might mitigate the harms of rampant individualism.

Not so many years ago, civil libertarians were angry about the imposition of seat belt usage in cars. The individual freedoms of all would be curtailed by seat belts in favour of a minority. Legislation was passed and before long the benefits of seat belts were readily apparent to all. The fuss about civil liberties died down and has not been heard since on that subject. This example illustrates that populations can be induced to change their behaviour even when the likely benefit to the individual is small.

It would be easy to make a list of behaviours that if changed would be likely to benefit health and the environment in the long term. Decreased speed limits for motor vehicles, decreased consumption of meat, decreased consumption of 'foreign foods', increased sharing of resources. Years ago the Maharishi Yogi said that if ten per cent of the world did transcendental meditation a new world order of peace and love would emerge. Peter Singer has said much the same about ethical living. Whether these ideas have any foundation is uncertain. What is certain is that changes in thinking, feeling and behaving by each and every one of us in relation to our ecologies and each other are needed to make the substantial changes that will be required. We will not voluntarily become ecological citizens until our identities are such. Our identities will not be ecological unless and until we act in ecological ways in our private and public lives day after day after day.

What resources do we have to bring about change? At first sight, very little in comparison with our economic masters. Our main resource is our children. They are our greatest gift and we owe them our greatest efforts. They are the human beings we are most likely to influence and our overwhelming responsibility is to them. Let us influence them ourselves and by reform of our educational system. If we raise our children so that their identities are intrinsically ecological and caring of all life, in a generation or two the necessary changes will have occurred.

The Green movement, leadership and health

We have frequently referred to the Green movement, conservationists and the Green agenda. We made the statement that Save the Bush is a health message. The Green movement is of great interest to us because, probably unbeknown to most Greenies, their agenda is part of an important international health movement that advocates extensive reforms, indeed non-violent revolution, and therefore attracts much antagonism.

Charles Rubin is a professor of political science at Dequesne University. His book *The Green Crusade* is a detailed critique of a series of populist writers who are environmentalists, such as Carson, Commoner, Ehrlich, Schumacher, and more recently the proponents of deep ecology.[9] What Rubin says about the frequent lack of scientific backing for many of the claims and predictions made by environmentalists is interesting. He maintains that the environmentalists' strength lies mainly in their ability to communicate and to motivate the general population. In a learned and detailed expose, he draws together various predictions of gloom and doom and suggests that a fundamental flaw in environmentalism is its reliance on the extra dimension: the value system or non-scientific component of our existence. We disagree with Rubin's conclusion. This component can be regarded as a strength of environmentalism. Rubin's main concern is the social and political implications; he sees all environmental actions ending in totalitarianism. If we are to stop the population increase we will need rules – who will promulgate the rules, who will enforce them? He believes that environmental actions will lead to attacks on individual freedom, democracy and liberalism.

Relying on the same facts Rubin uses, we would like to put a simple alternative viewpoint. The environmental movement is searching for alternative social and political systems because the present system palpably fails to deliver environmental health and human needs. The worldwide environmental situation is deteriorating and escaping from the bounds of possible control. This has forced the environmental movement into the political arena with the creation of Green parties. Although Green politics is fluid and disorganised, with a multitude of conflicting thoughts, Charles Rubin always sees the seeds of totalitarianism. Our feeling is that a form of totalitarianism has already arisen by an alternative mechanism. It is a totalitarianism hidden in the subtleties of the consumer society. Do we truly have the freedoms that Rubin cherishes when we wish to change the actions of the government or of a multinational company? We oppose totalitarianism since we believe it is the antithesis to ecological sustainability.

The importance of the Green movement is that it utilises community democracy and action to attain specific goals. This offers a vital alternative to a population directed by international organisations and companies. Studies of the beliefs of environmentalists show them to have views on social, economic and ethical issues which are radical and favour change. This sort of activism carries important messages for our future health. The delivery of health to much of the world, and to the developing world in particular, needs active non-government organisations (NGOs). The environmental groups have immense experience in these matters, delivering community-based solutions with minuscule budgets and unbridled enthusiasm. They therefore provide a tested community model to pursue environmental and health issues together.

There are other community-based organisations which are recognising their obligations to speak out on matters of equity, employment, health and the environment. Consider the Christian churches, which in Australia have expressed increasing alarm about unemployment levels, Aboriginal issues and the widening gulf between rich and poor. In poorer countries, the churches are often in the position of providing a nucleus for community activity. In *A Truly Civil Society*, Eva Cox has called this human, community activity 'social capital', for it offers the only viable alternative to the money capital that washes around the world's markets and determines the ruling structure of our society.[10] It is the community itself that will provide the social capital for environmental and health care and for social and economic support.

Communities working in these fundamental ways have faced the realisation that development of resources is a myth for well-being and health. In an assessment of natural and human resources made by the World Bank, Australia is the wealthiest country on Earth, yet the poor get poorer, the aged and sick have their share of this wealth and their entitlements cut, and the health of our indigenous people is appalling. Wealth does not translate into health, equity or care for the environment. There are many examples of countries discovering resources of immense value, yet obtaining no long-lasting social, community or other gain. This has applied particularly to the discovery of oil. The grass roots – the Green and socially oriented society – recognises that the solutions lie with society itself and not with capitalist or government promises of wealth and jobs.

Our relationship to the planet

We have predicted the likely course of Western scientific medicine – a minority will thrive on it but the health and well-being of the majority of the human race will languish because of damage to the Earth's environment. If we could make one change to the lives of individuals, it would be an attitudinal change, a change in values. We are deluding ourselves if we feel we can change the attitude to health imbued in the powerful banker in Sydney, automobile king in Germany or computer whiz in the USA. Health to them is a prescribed daily gym session and day off for golf, regular screening of weight, blood pressure and cholesterol, and artery and even organ replacement at the appropriate time to prolong their valued lives. These costs of scientific medicine are paid for by the company in salary packages and ordinary people indirectly pay for their medical care when they buy or use their products. The companies of these affluent executives pay little towards worldwide health and well-being for they move money around and pay few taxes; however they do lecture and influence governments on pay restraint and they finance some projects that do environmental harm!

As we have seen, health to the indigenous peoples of the forests of South-East Asia is the retention of their land and rivers for subsistence and their proximity to their ancestors who occupied the land for generations. They have health and well-being within their own definition although they will not live as long as the three Western executives. However, they will lose their health and well-being when their land is taken over for commercial exploitation in the name of progress and with the promise of jobs and consumer goods for all. The indigenous people will be relocated 'in their own interests'. These two value systems cannot be reconciled under present day governance.

We have attempted to draw many important issues together.

We have used scientific thought whilst recognising the drawbacks of science. We have laid emphasis on health and well-being, for each human being sees this as a fundamental issue which relates to themselves and their families and which they sometimes realise is not necessarily related to economic success and development. We also recognise the extra dimension in our relationship to this planet that cannot be based on 'growth', 'expansion', 'economic success' and so on.

An underlying theme in our thoughts has been one of conversion. Not the blinding light of religious revelation but the conversion that comes from careful thought and consideration. This conversion is not based entirely on thought and reason; it has to be based on instinct and our recognition of our place in nature. It is not possible for the hypothetical banker in Sydney to recognise this relationship without a childhood or adulthood wilderness experience to provide a realisation of fundamental issues. From this experience, which cannot be explained scientifically, flows a series of conclusions that affect our recognition of our place on Earth, our duty to others, and our innate health and well-being now and in the future. As scientists and philosophers, we ask you to consider them.

An important theme of our concern has been that humans, with their intelligence, inventiveness and acquisitiveness, have broken away from their biological and ecological need to be part of the total community of life. In so doing, they have overwhelmed the adaptive responses of their own bodies and minds as evidenced by the diseases of Western communities. While Western scientific medicine will 'fix' some of these problems, many others are likely to arise as the pace of change accelerates. We have made unlimited technological progress, yet our social structures remain primitive in that we are unable to cope with conflict, war and inequity, all of which carry dangers for everyone. We are also in the process of severing our links to the ecological web of life and this will have global consequences for our future health and well-being. Our simple message has been that to remain healthy we have to remain part of the web of life. Everything on the planet is interconnected and to function properly our culture, education, government and economic system must be unified, not compartmentalised, and we must recognise that every decision has environmental and health connotations.

Human intelligence and inventiveness, which know no bounds in the world of technology, can be harnessed to examine our predicament and to evolve new structures to lead us away from the inevitable destruction that will result from everlasting 'growth'. We speculate that these ideas and solutions will not evolve from more science and technology, but perhaps from the philosophers, the religious, the humanists or the environmentalists. In recent years there has perhaps been one text pointing the way: James Lovelock's *Gaia: A New Look at Life on Earth* was written not in a university or institute but in a house set in the English countryside.[11] Its creation reminds one of Darwin's deliberations a century before. Human health, like ecology, is a web of life-factors which interact to produce good health and well-being. Disrupt one or more of them and disease may result. Each of us, in health and disease, is more than the sum of our organs and systems. Each of us functions as a whole and medical practice ignores this at its peril. But our 'wholeness' includes our biological and spiritual attachment to the environment, to our locale or 'place', to other living things, and to a recognition of our place on the planet.

---

1. Aitlon Krenak, Co-ordinator of Indian Nation's Union, at the World Commission on Environment and Development Public Hearing.

2. Al Gore, *Earth in the Balance* (Houghton Mifflin, New York, 1992).

3. Peter Singer, *How Are We to Live?* (Text Publishing, Melbourne, 1993).

4. Theodore Roszak, *The Voice of the Earth* (Simon & Schuster, New York, 1992).

5. E.O. Wilson, *The Diversity of Life* (Belknar, Harvard, 1992).

6. David Smith, 'How Will It All End?' in *Health and the Environment*, ed. Bryan Cartledge (Oxford University Press, Oxford, 1994).

7. David Orr, What Is Education For? *Trumpeter* 8:3 (Summer 1991), pp. 99–102.

8. Robyn Eckersley, *Environmentalism and Political Theory: towards an ecocentric approach* (University College of London Press, 1992).

9. Charles T. Rubin, *The Green Crusade: Rethinking Routes of Environmentalism* (Free Press, New York, 1994).

10. Eva Cox, *A Truly Civil Society* (ABC Books, Sydney, 1995).

11. James Lovelock, *Gaia: A New Look at Life on Earth* (Oxford University Press, Oxford, 1979).

Acknowledgements

The authors wish to thank Matt Gaughwin and Philip Shearman for their advice and expertise, Roger Porter, Marjory Heath, Suzanne Heath and others who do not wish to be mentioned, for their support and for reading drafts of the book. We also thank Sue Suter and Clare Shearman for their secretarial help and our many academic collegues who have donated countless hours discussing with us the ideas presented here. Our thanks also to editor Amanda Ryder for her expertise.

Index

Last, J.M., 54
Lavers, P.J., 145
lead, 22, 148
Leakey, Richard, 53, 74
leptospirosis, 5
leukaemias, 6, 74
Liberal Parties, 164
liberal political philosophy, 163–171, 258
life expectancy, 11, 23
Lincoln, Abraham, 220
Lister, Joseph, 39
liver,
    human, 47
    transplant, 48
Livingston, John, 72
Los Angeles, and fortified living compounds, 142
Lovelock, James, 46, 116, 267
Lovins, L.H., , 25
Lowe, Ian, 207
Lyme disease, 67
lymphoma, 148

M
Mabo, 18
Mad Cow Disease, 43–45, 193
Mahler, H, 13
malaria, 25, 74, 80, 82, 96, 97–99, 182, 193
    cerebral, 97, 237
    description by Hippocrates, 98
    East Africa, 98
    Irian Jaya, 82
    Madagascar, 98
    Papua New Guinea, 97, 98
    spread to Australia, 98
Malaysia, 18
    Prime Minister, 166
malnutrition, 17, 143, 253
Malthus Thomas, 182

Mao, Chairman, 106
Maori, concept of health, 15, 16, 19, 53
Marine Mammal Protection Act, 148
Marx, and Marxism, 217–218, 224
mass extinction of species, 51, 53
McMichael, Tony, 119
measles, 165, 239
media, 108, 151, 168, 202, 211, 249, 257, 258, 259
medical science, 6
Mediterranean
    civilisations, 53
    Sea, 4, 98, 123
melanoma, 29
meningitis, 40, 231
mental health and ecology, 240–242, 250
mesothelioma, 21, 24
methane, 84
mice, 1, 67
mining, 144, 146
Ministry of Truth, 184
mist drip, 61
mobile phones, 79
molecular biology, 50
money, 129, 219–222
monoculture, 74, 243
Montreal Protocol (1987), 28
mosquitoes, 25, 96, 239
multinational companies, 87, 206
Murray-Darling river system, 173–174
Myers, Norman, 51

N
Nader, Ralph, 144, 159
NAFTA, 147
National Biodiversity Council, 72
    Competition Policy, 133
    Health and Medical Research
        Council, 10, 246, 252, 254–255